Science and Nonbelief

**Recent Titles in
Greenwood Guides to Science and Religion**

Science and Nonbelief

TANER EDIS

Greenwood Guides to Science and Religion
Richard Olson, Series Editor

Greenwood Press
Westport, Connecticut • London

Library of Congress Cataloging-in-Publication Data

Edis, Taner, 1967–
    Science and nonbelief / Taner Edis.
        p. cm.—(Greenwood guides to science and religion)
    Includes bibliographical references and index.
    ISBN 0–313–33078–6 (alk. paper)
    1. Science—History—Philosophy. 2. Religion and science—History.
3. Science and spiritualism—History. 4. Faith and reason—History.
5. Naturalism—History. 6. Nature—Religious aspects. I. Title.
II. Series.
    Q174.8.E35   2006
    501—dc22   2005020933

British Library Cataloguing in Publication Data is available.

Library of Congress Catalog Card Number: 2005020933
ISBN: 0–313–33078–6

First published in 2006

Greenwood Press, 88 Post Road West, Westport, CT 06881
An imprint of Greenwood Publishing Group, Inc.
www.greenwood.com

Printed in the United States of America

The paper used in this book complies with the
Permanent Paper Standard issued by the National
Information Standards Organization (Z39.48–1984).

10 9 8 7 6 5 4 3 2 1

Contents

Contents

Series Foreword

For nearly 2,500 years, some conservative members of societies have expressed concern about the activities of those who sought to find a naturalistic explanation for natural phenomena. In 429 B.C.E., for example, the comic playwright Aristophanes parodied Socrates as someone who studied the phenomena of the atmosphere, turning the awe-inspiring thunder that had seemed to express the wrath of Zeus into nothing but the farting of the clouds. Such actions, Aristophanes argued, were blasphemous and would undermine all tradition, law, and custom. Among early Christian spokespersons, there were some, such as Tertullian, who also criticized those who sought to understand the natural world on the grounds that they "persist in applying their studies to a vain purpose, since they indulge their curiosity on natural objects, which they ought rather [direct] to their Creator and Governor."[1]

In the twentieth century, though a general distrust of science persisted among some conservative groups, the most intense opposition was reserved for the theory of evolution by natural selection. Typical of extreme anti-evolution comments is the following opinion offered by Judge Braswell Dean of the Georgia Court of Appeals: "This monkey mythology of Darwin is the cause of permissiveness, promiscuity, pills, prophylactics, perversions, pregnancies, abortions, pornography, pollution, poisoning, and proliferation of crimes of all types."[2]

It can hardly be surprising that those committed to the study of natural phenomena responded to their denigrators in kind, accusing them of willful ignorance and of repressive behavior. Thus, when Galileo Galilei was warned against holding and teaching the Copernican system of astronomy as true, he wielded his brilliantly ironic pen and threw down a gauntlet to religious authorities in an introductory letter, "To the Discerning Reader," at the beginning of his great *Dialogue Concerning the Two Chief World Systems*:

Several years Ago there was published in Rome a salutory edict which, in order to obviate the dangerous tendencies of our age, imposed a seasonable silence upon the Pythagorean [and Copernican] opinion that the earth moves. There were those who impudently asserted that this decree had its origin, not in judicious inquiry, but in passion none too well informed. Complaints were to be heard that advisors who were totally unskilled at astronomical observations ought not to clip the wings of reflective intellects by means of rash prohibitions.

Upon hearing such carping insolence, my zeal could not be contained.[3]

No contemporary discerning reader could have missed Galileo's anger and disdain for those he considered enemies of free scientific inquiry.

Even more bitter than Galileo was Thomas Henry Huxley, often known as "Darwin's bulldog." In 1860, after a famous confrontation with the Anglican bishop Samuel Wilberforce, Huxley bemoaned the persecution suffered by many natural philosophers, but then he reflected that the scientists were exacting their revenge:

Extinguished theologians lie about the cradle of every science as the strangled snakes beside that of Hercules; and history records that whenever science and orthodoxy have been fairly opposed, the latter has been forced to retire from the lists, bleeding and crushed, if not annihilated; scotched if not slain.[4]

The impression left, considering these colorful complaints from both sides is that science and religion must continually be at war with one another. That view was reinforced by Andrew Dickson White's *A History of the Warfare of Science with Theology in Christendom*, which has seldom been out of print since it was published as a two-volume work in 1896. White's views have shaped the lay understanding of science and religion interactions for more than a century, but recent and more careful scholarship has shown that confrontational stances do not represent the views of the overwhelming majority of either scientific investigators or religious figures throughout history.

One response among those who have wished to deny that conflict constitutes the most frequent relationship between science and religion is to claim that they cannot be in conflict because they address completely different human needs and therefore have nothing to do with one another. This was the position of Immanuel Kant who insisted that the world of natural phenomena, with its dependence on deterministic causality, is fundamentally disjoint from the noumenal world of human choice and morality, which constitutes the domain of religion. Much more recently, it was the position taken by Stephen Jay Gould in *Rocks of Ages: Science and Religion in the Fullness of Life*:

I . . . do not understand why the two enterprises should experience any conflict. Science tries to document the factual character of the natural world and to develop theories that coordinate and explain these facts. Religion, on the other hand, operates in the equally important, but utterly different realm of human purposes, meanings, and values.[5]

In order to capture the disjunction between science and religion, Gould enunciates a principle of "Non-overlapping magisteria," which he identifies as "a principle of respectful noninterference."[6]

In spite of the intense desire of those who wish to isolate science and religion from one another in order to protect the autonomy of one, the other, or both, there are many reasons to believe that this is ultimately an impossible task. One of the central questions addressed by many religions is the relationship between members of the human community and the natural world. This is a central question addressed in Genesis, for example. Any attempt to relate human and natural existence depends heavily on the understanding of nature that exists within a culture. So where nature is studied through scientific methods, scientific knowledge is unavoidably incorporated into religious thought. The need to interpret Genesis in terms of the dominant understandings of nature thus gave rise to a tradition of scientifically informed commentaries on the six days of creation, which constituted a major genre of Christian literature from the early days of Christianity through the Renaissance.

It is also widely understood that in relatively simple cultures—even those of early urban centers—there is a low level of cultural specialization, so economic, religious, and knowledge-producing specialties are highly integrated. In Bronze-Age Mesopotamia, for example, agricultural activities were governed both by knowledge of the physical conditions necessary for successful farming and by religious rituals associated with plowing, planting, irrigating, and harvesting. Thus, religious practices and natural knowledge interacted in establishing the character and timing of farming activities.

Even in very complex industrial societies with high levels of specialization and division of labor, the various cultural specialties are never completely isolated from one another and they share many common values and assumptions. Given the linked nature of virtually all institutions in any culture, it is the case that when either religious or scientific institutions change substantially, those changes are likely to produce pressures for change in the other. It was probably true, for example, that the attempts of pre-Socratic investigators of nature, with their emphasis on uniformities in the natural world and apparent examples of events systematically directed toward particular ends, made it difficult to sustain beliefs in the old pantheon of human-like and fundamentally capricious Olympian gods. But it is equally true that the attempts to understand nature promoted a new notion of the divine—a notion that was both monotheistic and transcendent, rather than polytheistic and immanent—that focused on both justice and intellect rather than power and passion. Thus, early Greek natural philosophy undoubtedly played a role not simply in challenging but also in transforming Greek religious sensibilities.

Transforming pressures do not always run from scientific to religious domains, however. During the Renaissance, there was a dramatic change of thought among Christian intellectuals from one that focused on the contemplation of God's works to one that focused on the responsibility of the Christian to care for his fellow humans. The active life of service to humankind, rather than the contemplative life of reflection on God's character and works, now became the Christian ideal for many. As a consequence of this new focus on the active life, Renaissance intellectuals turned away from the then-dominant Aristotelian view of science, which saw the inability of theoretical sciences to change the world as a positive virtue. They replaced this understanding with a new view of natural knowledge, promoted in the writings of men such as Johann Andreae in Germany and Francis Bacon in England, which viewed natural knowledge as significant only because it gave humankind the ability to manipulate the world to improve the quality of life. Natural knowledge would henceforth be prized by many because it conferred power over the natural world. Modern science thus took on a distinctly utilitarian shape, a response due at least in part to religious changes.

Neither the conflict model nor the claim of disjunction, then, accurately reflect the often intense and frequently supportive interactions between religious institutions, practices, ideas, and attitudes on the one hand, and scientific institutions, practices, ideas, and attitudes on the other. Without denying the existence of tensions, the primary goal of this series is to explore the vast domain of mutually supportive and/or transformative interactions between scientific institutions, practices, and knowledge and religious institutions, practices, and beliefs. A second goal is to offer the opportunity to make comparisons across space, time, and cultural configuration. The series will cover the entire globe, most major faith traditions, hunter–gatherer societies in Africa and Oceania as well as advanced industrial societies in the West, and the span of time from classical antiquity to the present. Each volume will focus on a particular cultural tradition, faith community, time period, or scientific domain, so that each reader can enter the fascinating story of interactions between science and religion from a familiar perspective. Furthermore, each volume will include not only a substantial narrative or interpretive core, but also a set of primary documents which will allow the reader to explore relevant evidence, an extensive annotated bibliography to lead the curious to reliable scholarship on the topic, and a chronology of events to help the reader keep track of the sequence of events involved and to relate them to major social and political occurrences.

So far I have used the words "science" and "religion" as if everyone knows and agrees about their meaning and as if they were equally appropriately applied across place and time. Neither of these assumptions is true. Science and religion are modern terms that reflect the way that we in the industrialized West organize our conceptual lives. Even in the modern West, what we mean by science and religion is likely to depend on our political orientation, our scholarly background, and the faith community to which we belong. Thus, for example, Marxists and Socialists tend to focus on the application of natural knowledge as the key element in defining science. According to the British Marxist scholar Benjamin Farrington, "Science is the system of behavior by which man has acquired mastery of his environment. It has its origins in techniques . . . in various activities by which man keeps body and soul together. Its source is experience, its aims, practical, its *only* test, that it works."[7] Many of those who study natural knowledge in pre-industrial societies are also primarily interested in knowledge as it is used and are relatively open regarding the kind of entities posited by the developers of culturally specific natural knowledge systems or "local sciences." Thus, in his *Zapotec Science: Farming and Food in the Northern Sierra of Oaxaca*, Roberto González insists that

Zapotec farmers . . . certainly practice science, as does any society whose members engage in subsistence activities. They hypothesize, they model problems, they experiment, they measure results, and they distribute knowledge among peers and to younger generations. But they typically proceed from markedly different premises—that is, from different conceptual bases—than their counterparts in industrialized societies.[8]

Among the "different premises" is the Zapotec scientists' presumption that unobservable spirit entities play a significant role in natural phenomena.

Those more committed to liberal pluralist society and to what anthropologists like González are inclined to identify as "cosmopolitan science," tend to focus on science as a source of objective or disinterested knowledge, disconnected from its uses. Moreover, they generally reject the positing of unobservable entities, which they characterize as "supernatural." Thus, in an *Amicus Curiae* brief filed in connection with the 1986 Supreme Court case that tested Louisiana's law requiring the teaching of creation science along with evolution, *72 Nobel Laureates, 17 State Academies of Science and Seven Other Scientific Organizations* argued that

Science is devoted to formulating and testing naturalistic explanations for natural phenomena. It is a process for systematically collecting and recording data about the physical world, then categorizing and studying the collected data in an effort to infer the principles of nature that best explain the observed phenomena. Science is not equipped to evaluate supernatural explanations for our observations; without passing judgement on the truth or falsity of supernatural explanations, science leaves their consideration to the domain of religious faith.[9]

No reference whatsoever to uses appears in this definition. And its specific unwillingness to admit speculation regarding supernatural entities into science reflects a society in which cultural specialization has proceeded much further than in the village farming communities of southern Mexico.

In a similar way, secular anthropologists and sociologists are inclined to define the key features of religion in a very different way than members of modern Christian faith communities. Anthropologists and sociologists focus on communal rituals and practices that accompany major collective and individual events: plowing, planting, harvesting, threshing, hunting, preparation for war (or peace), birth, the achievement of manhood or womanhood, marriage (in many cultures), childbirth, and death. Moreover, they tend to see the intensification of social cohesion as major consequence of religious practices. Many Christians, on the other hand, view the primary goal of their religion as personal salvation, viewing society at best as a supportive structure and at worst as a distraction from their own private spiritual quest.

Thus, science and religion are far from uniformly understood. Moreover, they are modern Western constructs or categories whose applicability to the temporal and spatial "other" must always be justified and must furthermore be interpreted as the means by which we organize our understanding of the actions and beliefs of people who would not have used those terms themselves. Nonetheless it does seem to us not simply permissible but probably necessary to use these categories at the start of any attempt to understand how actors from other times and places interacted with the natural world and with their fellow humans. It may ultimately be possible for historians and anthropologists to understand the practices of persons distant in time and/or space in terms that those persons might use. But that process must begin by likening the actions of others to those that we understand from our own experience, even if the likenesses are inexact and in need of qualification.

The editors of this series have not imposed any particular definition of science or of religion on the authors, expecting that each author will develop either explicit or implicit definitions that are appropriate to their own scholarly approaches and to the topics that they have been assigned to cover.

Richard Olson
Claremont, California

NOTES

1. Tertullian, 1896–1903, "Ad nationes," *The Ante-Nicene Fathers,* ed. Alexander Roberts and James Donaldson, trans. Peter Holmes (New York: Scribner, 1900), 3: 133.

2. Christopher Toumey, *God's Own Scientists: Creationists in a Secular World* (New Brunswick, NJ: Rutgers University Press, 1994), 94.

3. Galileo Galilei, *Dialogue Concerning the Two Chief World Systems: Ptolemaic and Copernican* (Berkeley: University of California Press, 1953), 5.

4. James R. Moore, *The Post-Darwinian Controversies: A Study of the Protestant Struggle to Come to Terms with Darwin in Great Britain and America, 1870–1900* (Cambridge: Cambridge University Press, 1979), 60.

5. Stephen Jay Gould, *Rocks of Ages: Science and Religion in the Fullness of Life* (New York: Ballantine, 1999), 4.

6. Ibid., 5.

7. Benjamin Farrington, *Greek Science* (Baltimore: Penguin, 1953).

8. Roberto González, *Zapotec Science: Farming and Food in the Northern Sierra of Oaxaca* (Austin: University of Texas Press, 2001), 3.

9. *72 Nobel Laureates, 17 State Academies of Science and Seven Other Scientific Organizations. Amicus Curiae* brief in support of Appelles Don Aguilard, et al. versus Edwin Edwards in his official capacity as Governor of Louisiana et al. (1986), 24.

Preface

A book on science and nonbelief has to approach its subject from many different perspectives. On the one hand, there are intellectual debates about what modern science says about claims of supernatural and transcendent realities. On the other hand, science and religion are both significant social institutions, so their relationship is always complicated by broad political considerations. Thus, in these pages I touch on social as well as natural science, discuss philosophical disputes alongside scientific ideas, and make sure I convey enough of the complex historical interactions between science and nonbelief.

No one is an expert on everything, and my approach is definitely influenced by my background as a physicist—albeit one who is fascinated by varieties of creationism and other attempts to blend the supernatural with modern science. In addition, a manageable book must be selective. I have not covered the broad sweep of possible positions involving science and nonbelief, concentrating instead on what I call "science-minded nonbelief," which takes the naturalism of current science as the leading reason to reject the existence of spiritual realities. And my comments on the political and social context concerning science and nonbelief emphasize conditions in the United States.

Nonetheless, I cover a number of different subjects and aim to do it in a readable manner. So I relied on a group of "previewers" of my manuscript as it developed. They have been very helpful in pointing out everything from blatant errors to unclear passages. All mistakes remaining in the book (and there are bound to be some) are mine. I would like to thank Amy Sue Bix, Richard Carrier, Eva Durant, Mike Huben, Gert Korthof, Jim Lippard, William D. Loughman, Jeffery Jay Lowder, Richard G. Olson, Markus Pössel, Ilkka Pyysiäinen, Eleanor Schechter, Brad Smith, and Matt Young for all their help.

I also should thank Kevin Downing of Greenwood Press, who had to put up with my floundering in matters such as finding illustrations, and Barry Karr and Tom Flynn, who were very helpful when I was looking for permission to use material from the *Skeptical Inquirer* and *Free Inquiry*.

Chronology of Events

585 B.C.E.	Ionian philosopher Thales said to have predicted an eclipse.
384–322 B.C.E.	Life of Aristotle, Greek philosopher and scientist.
c. 50 B.C.E.	Lucretius writes *De Rerum Natura* (*On the Nature of Things*).
c. 100–178 C.E.	Life of Ptolemy, Greek astronomer and astrologer.
866	Death of Al-Kindi, influential Mutazilite (rationalist) Muslim scholar.
1265	Thomas Aquinas begins writing his *Summa Theologica*.
1541	Death of Paracelsus, physician and occultist.
1543	Death of Nicolas Copernicus, and publication of his *De Revolutionibus*, arguing that the earth revolved around the sun.
1600	Giordano Bruno executed, condemned for heresy by the Inquisition.
1633	Trial of Galileo Galilei for his support of Copernican theory.

1641	René Descartes's *Meditations on First Philosophy* presents and defends dualism.
1687	Isaac Newton's *Principia* published.
1734–1815	Life of Anton Mesmer, founder of Mesmerism.
1751–1777	Publication of the *Encyclopédie* of Diderot and d'Alembert.
1779	Posthumous publication of David Hume's *Dialogues Concerning Natural Religion*.
1799	Laplace's *Celestial Mechanics* published.
1802	William Paley's *Natural Theology* published, with "watchmaker" argument pointing to a designer-God.
1848	Karl Marx and Friedrich Engels's *Communist Manifesto*.
1848	The Fox sisters in New York, instrumental in starting Spiritualist movement.
1855	First edition of Ludwig Büchner's *Force and Matter*.
1859	Charles Darwin's *Origin of Species* published.
1865–1899	Robert Green Ingersoll tours the United States, speaking on subjects such as freethought.
1907–1916	Albert Einstein develops general relativity, his theory of gravity.
1924–1928	Most intense development of the basic concepts of quantum mechanics.
1927	Publication of Sigmund Freud's *The Future of an Illusion*.
1927–1933	Georges-Henri Lemaître works on the earliest version of the big bang theory.
1935	J. B. Rhine founds the Parapsychology Laboratory at Duke University.
1937–1942	The Atanasoff-Berry computer, the world's first digital computer.

1947	Kenneth Arnold's UFO sighting and the beginning of post–World War II popularity of flying saucers.
1947	Alleged flying saucer crash at Roswell, New Mexico.
1953	Discovery of DNA by James Watson and Francis Crick.
1961	Henry M. Morris and John C. Whitcomb publish *The Genesis Flood* and revive young-earth creationism in the United States.
1973	First Templeton Foundation Prize for Progress in Religion presented.
1976	Committee for the Scientific Investigation of Claims of the Paranormal started.
1990	The Hubble Space Telescope launched.
1990	Beginning of the Human Genome Project.
1990–1999	Decade of the Brain, the accelerated development of neuroscience.
1991	With Phillip Johnson's *Darwin on Trial*, intelligent design creationism becomes visible.
1996	The James Randi Educational Foundation is founded, offering a $1 million prize for demonstration of a paranormal phenomenon.

Chapter 1

Science, Philosophy, and Religious Doubt

PHILOSOPHERS, DOUBTING AND DEVOUT

There have always been nonbelievers. They have been rare—most people accept their local religion; human societies usually define themselves through supernatural beliefs. Even so, from crusty peasants who think their priest speaks nonsense to philosophers who think only the unsophisticated take stories of the gods at face value, religion invites dissenters. They need not profess nonbelief; in fact, they are more often heretics who favor different gods, such as the early Christians who were called atheists because they disbelieved the established Roman divinities. We rarely hear of skeptics who go so far as to reject all gods and demons, and when we do, it is hard to say if the accusation of complete infidelity is true or an exaggerated insult by heretic hunters. Nevertheless, there must have been a few nonbelievers even thousands of years ago: skeptics who thought that this world is all there is, that the gods are phantoms, and that magic is only illusion.

And there have always been the curious, the tinkerers. Though science as we understand it today has not been around longer than a few centuries, philosophical speculation about how the world works has been common enough, not to mention the local wisdom about which herbs can heal. And when some people begin to draw pictures of the world independent of religious beliefs, interesting things happen. Some respond by harmonizing religious and secular knowledge, trusting that human reason can only deepen how we understand divine purposes. Some hold fast to the beliefs they were handed down, treating mere worldly wisdom with suspicion. Yet others become skeptical, letting reason lead to more and more nonbelief.

So early on, if the protoscientific knowledge and practices achieved by ancient cultures had to do with doubting the gods at all, this doubt came about as part of a more general tension between philosophy and religion, between reason and revelation. People with a philosophical turn of mind valued independent reasoning and critical thought, while religion relied on authoritative communications from higher realities transcending the everyday world. Particular notions of supernatural realities drew support from mystical practices, from religious specialists who appeared to commune successfully with divine powers, and from the way tight-knit communities wove their moral sanctions together with their perceptions of divinity. Religious innovation was suspect; even when a prophet succeeded in changing minds, this usually split off a new community that was devoted to preserving the new revelation from all criticism and corruption. A critical attitude does not mix well with such conservative devotion, so disputes between philosophers and priests were bound to flare up on occasion.

Few ancient societies produced philosophy as a sustained enterprise. A holy man's disciples would often preserve his teachings and further interpret them, embellishing his wisdom with miracle stories to demonstrate divine endorsement. The ancient Greeks are our best-known example of how *critical* thought emerged, when the successors of a noted thinker felt free to revise what the master had proposed. Thales, the earliest-known Ionian philosopher, is said to have argued that water was the primal substance from which all else was derived. Other Ionians would agree that there had to be a primal substance, but they criticized Thales, proposing candidates other than water (Russell 1935, chapter 2).

In a philosophical environment, criticism of religious beliefs can develop, particularly if social changes conspire to make the old doctrines no longer look obvious. As the world of the Greeks became richer and more cosmopolitan, the gods of Homer and Hesiod became hard to believe in literally. Educated denizens of antiquity, though few, were all too likely to be aware of other mythologies and different priesthoods. They had to wonder what really was true about the divine, noticing that people pray to different kinds of gods and interpret their spiritual experiences within conflicting schemes of ultimate reality.

Xenophanes, a near contemporary of Thales, observed that "if cattle or horses or lions had hands . . . horses would draw the forms of their gods like horses and cattle, like cattle" (Kirk and Raven 1962, 169). Religion was particular to cultures, but reason could reveal universal truths. To the sophisticated, philosophy looked like a better path than popular piety.

Some important themes in the philosophical criticism of religion emerged early. Popular beliefs about transcendent realities seemed metaphysically untidy. Philosophers are often tempted by armchair reasoning, attempting to construct grandiose systems using nothing but "pure reason" and maybe some general facts no sane person disputes. Many ancient philosophers were attracted to Platonic ideas, which portrayed the messy, impermanent, and imperfect world we live in as a reflection of what was more rational, timeless, and perfect. The higher levels of reality beyond ours were not just literally more real, but also more spiritual, less entangled with mere matter. Though such a vision was certainly friendly to some sort of perfect God as the ultimate source of everything, it could also be religiously dangerous. The higher realities of the philosophers tended to become more abstract, impersonal, and distant—not the sort of things that could answer prayer, perform healing magic, or underwrite the authority of a priesthood. Moreover, philosophical realities could supposedly be demonstrated by reason, making prophets superfluous or at best useful messengers to the ignorant masses who could not appreciate the finer points of philosophy.

The later Neoplatonic philosophers and their heirs within Christianity and Islam would dispute many orthodox doctrines. Not only would they express skepticism about prophets, but they would find the miracles of popular religion vulgar and superstitious. They would deny that the world was created from nothing; instead, they would say all that exists was eternally emanated from a remote divine source, like light from the sun. And they would claim that humanlike purposes and passions and maybe even knowledge of earthly goings-on would only be imperfections impossible in an eternally changeless God. Indirectly, such ideas had influence beyond the elites. In striving to become intellectually respectable, Christian thinkers adopted philosophical language and found themselves tangled in so-called heresies about the nature of God and the incarnation of Christ. Even today, they still struggle with the criticism that doctrines like the Trinity and Jesus being both God and man make no sense. Among Muslims, rationalists called Mutazilites trusted reason and Greek philosophy. They used allegories to try and explain away anthropomorphic passages in the scriptures, declared that the Quran was the created, noneternal speech of God, and offended the traditionalists in many other ways (Fakhry 2004). When the Mutazilites attained political power, they even conducted an inquisition against those who would dissent from their view of religion. Orthodox Muslims stamped out the Mutazilites, and still remain deeply suspicious of those who want to elevate reason alongside revelation.

Another important strain of philosophical criticism of religion had a moral flavor. Traditionally, the most important reason to doubt has been the problem of evil. As the Greek philosopher Epicurus asked, Is God willing to prevent evil, but not able?—Or is God able, but not willing? Why, if the divinity is supposed to be all that is perfect, does it create and rule over such a nasty place as our world? The religious have devised many ways around this difficulty. The priests learned to promise all would be set aright in a heavenly afterlife into which only the devout would be allowed. Philosophers got into the act as well; many Platonists, for example, were partial to the notion that evil is only nonbeing, a deprivation due to our distance from the perfection that was God. While such proposed solutions swept some intellectual worries under the carpet, they rarely impressed those seeking relief from the nonbeing by which they were afflicted. And for anyone inclined toward nonbelief, they looked more like excuses than real solutions.

Today, doubters still ask why we should believe a perfect God created our world with all its evils. Freethinkers proudly champion reason over revelation, laying claim to the philosophical heritage of humanity. However, the historical conflict between reason and revelation is easy to exaggerate. The result of engaging in philosophy has more often been a more sophisticated, cleaned-up, retooled religious belief rather than nonbelief. Few philosophers before modern times publicly argued that there were no nonnatural spiritual realities of any sort; even those accused of complete nonbelief were likely skeptical only of popular religion or were labeled as nonbelievers by opponents. For example, al-Rawandi, the "Great Infidel" among Muslim philosophers, seems to have rejected Islam and prophetic religion. Although he is claimed as a hero by some modern dissidents from Islam (Ibn Warraq 1995, 259–260), it is much harder to say that he disbelieved in the God of the Greek philosophical tradition.

One reason philosophers usually managed to reconcile reason and revelation was that though well-constructed arguments have an intrinsic intellectual appeal, philosophical disputes do not take place in a vacuum. An intellectual who devoutly believes that the religion of the emperor is the one true faith and is skilled in arguing so is more likely to have a successful career than a skeptic. He will put more effort into explaining religious diversity as inferior but valuable glimpses into the transcendent granted to all peoples. His books will be more likely to be copied rather than burnt. Aristotle's works would have had an even tougher time surviving to this day if he had not acknowledged some version of a God and claimed to demonstrate its existence.

Until recently, then, full-blown nonbelief was very rare in intellectual history. It remained a potentiality in critical philosophical thought, and it might have occasionally emerged in individuals, but if so, it was dangerous to express in public and next to impossible to propagate as part of a developing intellectual stream. As James Thrower puts it, in the Greek tradition there are "very few examples of avowed and explicit atheism" (2000, 50). He adds, however, that "we find the beginnings of both the naturalistic and the sceptical outlook on life which will play so large a part subsequently in the development of a fuller and explicit atheism."

THE OLD SCIENCE AND THE OLD FAITH

Science could not have emerged without the critical reasoning skills cultivated by the philosophical tradition. Philosophy, however, is not enough. Curiosity about the natural world also helps, as does seeking systematic reality checks on theories. If premodern societies rarely sustained critical philosophy, they even more rarely produced scientific efforts that went beyond collecting immediately useful information about their environments. Intellectually and institutionally, what science there was could not be separated from religion and philosophy.

Nevertheless, with hindsight we can pick out occasions of early science, and the Greeks again figure prominently.[1] Pre-Socratic thinkers such as Thales appear to have conceived of the physical world as being governed by something like natural law rather than the personal whims of supernatural beings. We begin to see the notion of mathematically describable regularities: Thales is even reported to have predicted the eclipse of 585 B.C.E. Although Greek ideas about a lawful world were never fully naturalistic in the modern sense, they fed into currents like atomism, the claim that everything around us is composed of tiny atoms that interact through natural causes. For the history of nonbelief, Epicurean philosophy, as expounded by the Roman poet Lucretius Carus in the first century B.C.E. is most interesting. Lucretius proposes an atomistic physics to explain practically everything, including those phenomena taken as evidence for a spiritual view of the world, and deploys this materialism in service of a view that any gods are practically irrelevant to human affairs (Lucretius 1995).

Early science and religious doubt, however, were not too closely connected. Materialist ideas like those of Lucretius were never very convincing. His atomism was more metaphysical speculation than real physics, and his explanations became particularly strained when psychological matters came up. For example, Lucretius cheerfully accepted the existence of souls: they just happened to be composed of particularly fine, ethereal atoms. To avoid a completely predetermined world, he arbitrarily declared that his atoms could "swerve" haphazardly. The attraction of this style of philosophy was always its moral aspect and its claim to help adherents live a good life, not its speculative "science."

In fact, the best-developed aspects of Greek science were more likely to reinforce spiritual perceptions of the world. Greek science included some straightforward observation and classification, for example, in the biological work of Aristotle. The accomplishments of Archimedes, among others, illustrate how the Greeks developed some sound empirical laws and basic engineering knowledge. Their physics, however, remained a catalog of isolated facts—practical knowledge of little consequence for questions concerning the gods. The crowning glory of Greek science was their mathematics and astronomy. Here, not only do we find mathematical modeling and prediction, but ambitious theorizing that tries to make sense of the universe. Unlike earthbound physics and biology, Greek astronomical thinking went beyond collecting facts like stamps. Astronomy was an integral part of philosophical thought that associated the heavens with perfection. In the Aristotelian physics that remained intellectually dominant until only a few centuries ago, the sublunar realm we occupied was disorderly, imperfect, and subject to suffering and decay, while heavenly bodies partook in the changeless mathematical order of the divine.

So the pinnacle of Greek science had a distinctly astrological flavor.

For example, Ptolemy, the author of the *Almagest*, the most comprehensive astronomical treatise of the ancient world, also wrote the most
comprehensive astrological treatise, the *Tetrabiblos*. Ancient astrology
was not like the soup of New Age psychological metaphors or the
mindless newspaper love-prospects-for-Taurus columns common
today. On the scientific side, astrology was linked with the earth-
centered Ptolemaic picture of the universe, an impressive mathematical model by any standard. On the philosophical side, the
many-layered planetary spheres that made up the heavens (see Figure 1.1) were literally the abode of divine emanations manifesting
higher realities. Even popular religions drew on such philosophical
and scientific ideas, the best available at the time. Before Christianity
became its official faith, the Roman Empire was home to astral religions, serious rivals to Abrahamic monotheism. The heavenly bodies
were gods or angels. The soul's real home was the heavens, and astral religions promised ways to ascend to the realm of permanence
and join the immortals. The monotheistic religions themselves were
influenced by Greek astrology and cosmology. Christian and Muslim
holy writings preserve elements of both an older, traditional Near
Eastern cosmology with a flat earth and heaven above, and the conception of multiple Ptolemaic spheres. Medieval Muslim astronomy
was a continuation of the Greek science, including its astrological elements, but in a monotheistic context (Wright 2000).

Greek science and philosophy could influence religion because it
enjoyed prestige as part of the world picture of educated elites. Any
religion with aspirations beyond attracting lower-class followers
would have to accommodate philosophy. And so Saint Augustine, for
example, would advise his fellow Christians not to be too literal about
interpreting scripture, hoping to avoid the embarrassment of contra-
dicting the sophisticated philosophy of his time.

The most developed forms of the old science emerged in Islamic
lands. Muslims preserved and extended Greek work on mathematics,
astronomy, architecture, optics, and medicine, liberally laced with
what today we would consider rank superstition. But the few intellectual challenges to Islam came from the philosophical elements of
the intellectual tradition, not from the practical sciences. The knowledge of nature and speculative philosophy remained lesser "foreign
sciences" alongside greater sciences such as Quranic interpretation
and jurisprudence. In any case, all these enterprises were integrated
within an overall intellectual attitude that set God at the center, pre-

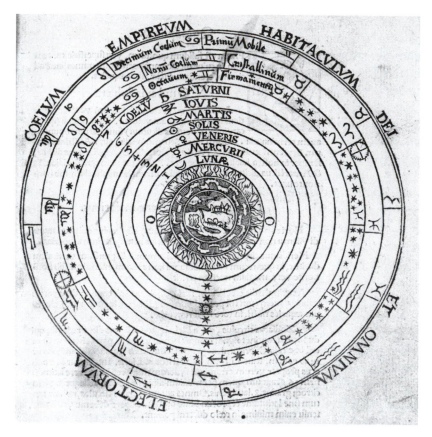

Figure 1.1 A simplified, medieval Christianized version of the Aristotelian universe. The earth is in the center of various planetary spheres, culminating in the motionless empyrean heaven, inhabited by God and all the elect. (Peter Apian, *Cosmographicus Liber*, 1524, col. 6. Courtesy The Lilly Library, Indiana University, Bloomington)

siding over all aspects of the universe, both hidden and knowable. Even today, prominent Muslim thinkers such as Seyyed Hossein Nasr denounce the erosion of this traditional view of intellectual life, hoping to return natural science to a position subordinate to an overall religious and metaphysical framework (Nasr 1989).

So the science of antiquity was not substantially associated with religious doubt. Even with hindsight, finding echoes of a conflict between science and religion is difficult. The precursors of natural science were usually little more than narrow areas of practical knowledge, with only a modest role in overall intellectual life. Where science was more ambitious, it also was spiritualized and absorbed into a religious perception of the world.

There were also few occasions for friction between scientists and religious institutions—there were no scientists as such. Scientific activity took place under the auspices of the philosophical tradition or within religious bodies themselves. By medieval times, the Islamic regions were intellectually more advanced, and philosophers with scientific interests supported themselves primarily as jurists and religious scholars. In Western Europe, education collapsed for a long while, though intellectual development continued in monasteries. Yet although Europe lagged far behind, modern science was to emerge there, often advanced by devoutly religious thinkers (Grant 2004).

THE EUROPEAN ENLIGHTENMENT

Western Europe in the eighteenth century had become a very different place than the fifteenth century, for ordinary people as well as elites. A revived rationalist philosophy made its mark, the common religious outlook of Catholic Christianity fragmented, new continents were discovered, and modern science emerged. Politics became increasingly democratic and nationalistic; capitalism arrived on the scene. The arts and music, technology, family structures—nothing, it seemed, could escape innovation, even revolution. Europe started to become modern.

Today's science and forms of nonbelief both derive from the great changes of the European Enlightenment. Though a modern scientist enjoys vastly advanced equipment and much more powerful theories, she has no difficulty seeing the kinship of her work with that of Galileo and Newton. And the arguments and attitudes common among today's rejecters of all things supernatural are recognizable in many Enlightenment thinkers.

For nonbelief, a very important development was the revival of classical philosophy. This is not to say that the Renaissance was a sudden awakening from the Dark Ages. Late medieval intellectual culture, which flourished inside the Catholic Church, was quite vibrant. Many of the themes that would find expression in later philosophy and science were already being discussed in the twelfth century. Still, these intellectual developments fully bubbled up with the Renaissance. In particular, intellectual life found channels outside of the church. A new set of more secular intellectuals emphasized those elements of classical philosophy that were more irredeemably pagan, not as easily absorbed into the world picture of the medieval church.

Lucretius, for example, was rediscovered, and became influential enough that orthodox thinkers had to attack his arguments. Astrology enjoyed a revival. Elite thinking did not reject the supernatural so much as drift away from the tightly organized, communal religion of medieval Catholicism. The new outlook was more individualist and more open to occult ideas. For example, Paracelsus, a physician and occult philosopher of the sixteenth century, was as representative of the innovative thinking of his time as figures who are more celebrated today, such as Galileo. Natural philosophy, which became science, very often included an interest in natural magic—it tried to investigate the fundamentally spiritual nature of our existence. Astronomers practiced astrology, and educated elites learned how the microcosm of human life was intimately connected to the macrocosm of the heavens.

With the fragmentation of religious authority, spiritual options proliferated. On the one hand, this meant wars of religion. On the other hand, intellectuals had to rely on reasoned arguments that commanded more than a narrowly sectarian appeal. Even as modern science was taking shape, religious and philosophical disputes increasingly took on a scientific flavor. The nature of our world was at issue.

The religiously orthodox were no less interested in harnessing new ideas about nature than the spiritual innovators. Hence, early on, a mechanistic philosophy became attractive to religious thinkers because of its promise to defeat more occult views. Mechanists conceived of the natural world as largely the domain of physics, where material objects interacted by strictly mechanical means. Slipping into the materialism of a Lucretius was no great danger, because mindless mechanical processes clearly did nothing to account for intelligence or any other spiritual, mind-related aspect of our existence. René Descartes, the best known of the mechanical philosophers, also contributed to mathematics, developed a vortex theory of planetary motion, and published proofs of God's existence. He argued that animals were mechanistic automatons, while humans were unique in possessing a rational soul—a mysterious thinking substance beyond any material mechanism.

So with the mechanical philosophy, mind and matter became separated by a deep gulf, and miracles became divine suspensions of otherwise inviolable laws of nature. This conception of the world, with sharply distinguished material and the spiritual realms, had no room for natural magic, for the gradual shading of the natural into the supernatural. Access to spiritual realities had to be channeled through the proper authorities rather than through astrologers and alchemists.

Naturally, when some intellectuals began to express nonbelief, they also looked to science to help make their case. Just a century after Descartes, philosophers of the French Enlightenment promoted skepticism toward supernatural religion. Some, like Voltaire, while severely anticlerical and dubious about miracles, were deists: they believed in a remote creator-God who did not interfere with the natural order. Others, such as Denis Diderot, abandoned belief in God. Science, for many Enlightenment thinkers, led toward a godless view of the world because it replaced divine purposes with natural causes. Moreover, reason could provide us with power to improve our lives and free us from fears of supernatural forces. Knowledge would set us free—from the gods, their priests, and the kings the ancient religions propped up. The new science promised replacing mystery with investigation, the unknowable purposes of God with a rational order. And so Diderot's greatest achievement was the monumental *Encyclopédie* he put together with Jean D'Alembert, a work devoted to gathering and disseminating the best of eighteenth-century knowledge about the world.

Other enduring themes of arguments that science supports nonbelief also emerged as Europe made its transition to modernity. La Mettrie, Diderot's contemporary, took the mechanical philosophy to its limits, portraying humans—intelligence, consciousness, and all—as machines in a universe of unbendable natural laws. In England, Thomas Hobbes proposed to explain religion itself in materialist, scientific terms, as something quite other than the human spirit's response to the divine. By today's standards, their views look overly ambitious, perhaps even crude, but it is significant that such ideas began to be expressed at all.

Even with science gaining in intellectual stature, however, nonbelief still received its strongest support from currents in philosophy. Theistic religion had long harnessed the philosophical tradition to its own ends, and alongside miracles and spiritual experiences, the strongest reasons it gave to believe in a God were ambitious metaphysical proofs of God. Such proofs were always a staple of philosophical theology, and were perfected in late medieval times. Three in particular stood out:

1. The ontological argument, which claims God's nonexistence is a logical impossibility;
2. The cosmological argument, which states that the universe as a whole requires a cause beyond nature; and
3. The argument from design, which explains the functional order found in the universe by attributing it to design by a supernatural intelligence (see Figure 1.2).

Though the cosmological and design arguments refer to features of the world not apprehended by pure reason alone, causality and order are very broad concepts. So these arguments floated in a rarefied metaphysical sphere, with no direct connection to scientific matters. Nevertheless, as the old cosmology fell apart, so did the confidence in traditional metaphysical reasoning. Even today, the philosophy of religion revolves around the big three proofs. Many introductory texts give their classic statements by medieval Catholic philosopher-priests such as Thomas Aquinas and Anselm, and then proceed to the proofs' Enlightenment critics, lingering especially on David Hume and Immanuel Kant. Ever since their demolition work, those who deny the force of the big three have had the upper hand. Hume and Kant removed the aura of rational certainty from God.

Where science was concerned, however, the effect of Hume's arguments in particular was ambiguous. Hume drew on the current in classical philosophy that was skeptical of the possibility of attaining knowledge. Skepticism about all knowledge may erode confidence in the dominant intellectual position—Christianity in Hume's case—but it can be deployed against much else. Modern philosophical defenders of God find much in Hume to help deflate science and isolate religion from criticism. Hume was also not very effective in refuting the argument from design, reaching no certain conclusion on the matter (Hume 1992; Rachels 1991). Though Hume undermined the more metaphysical version of the design argument, devout scientists increasingly argued that their work established the existence of a God. The world revealed by scientific investigation, they claimed, was a world exquisitely designed for human life. It was not a world that blind chance or mindless material forces could put together.

The relationship between science and religion during the European Enlightenment, then, was not simply that with the growth of science,

Figure 1.2 God designs the universe. The classic design argument attributes the intricate order in the universe to divine design. (From Österreichische Nationalbibliothek [Austrian National Library], Vienna, Latin MSS, MS. 2554, fol. 1r. Erich Lessing/Art Resources, NY)

religious belief faltered. Intellectual nonbelief was still driven primarily by philosophy, not science. And the new science also held out promise for religion: that this new, powerful way of generating knowledge could support traditional ideas about transcendent spiritual realities, and that this support would be *independent*—better than the arbitrary reliance on revelation and authority that rationalist philosophers rightly held in suspicion.

Nevertheless, the Enlightenment did produce a significant erosion of faith, especially among educated elites. The rise of science and non-belief were linked together, though often indirectly. The scientific revolution took place in a social environment where religious authority had weakened, and enterprises such as law and economic activities were established independently of the church (Huff 2003). Science, though starting in monasteries, could go its own way. The social fragmentation setting in through Europe created space for a science that was not constrained by religious doctrines. Both nonbelievers and early scientists benefited from this new freedom.

The early episodes of conflict between science and religion should also be seen in the light of science gaining independence as an institution. Today, nonbelievers and scientists alike celebrate martyrs to intellectual freedom such as Giordano Bruno, who proposed that the universe was infinite and ended up burned at the stake, or Galileo, who was forced by the Inquisition to recant his view that the Earth moved. Yet both were committed believers in God and other supernatural realities. The conflict was about which institutions would have authority in defining reality. As networks of thinkers working on scientific questions crystallized and began to form institutions, they naturally demanded autonomy in describing the natural world.

Intellectuals gaining independence from religious authority would not have meant much, if it were not for the powerful results that the new science produced. Physicists did not just propose new and improved conceptions of how bodies moved and how the universe was shaped; they made precise mathematical statements and tested them with careful experiments. The Greek physics and cosmology favored by the medieval church simply did not compare. Galileo and his fellow scientists began to reunite the heavens and the earth, and they were persuasive.

The towering figure of early modern science, who formulated the most enduring form of the new mathematical physics, was Isaac Newton. Though he is remembered for his physics—those principles of mechanics that physicists and engineers still use to describe everyday physics are known as "Newtonian"—only a portion of Newton's life's work was in science and mathematics. He was even more interested in alchemy and biblical prophecy. His physics, though, was what made a lasting impression. In the ancient conception, objects moved according to their various natures. So there was little to do but catalog what different types of object did, finding some vague patterns according to the various mixtures of air, fire, earth, and water from which they were supposed to be composed. In the Newtonian universe, the same fundamental laws of physics underlay the motions of *everything*, including planets and any and all earthly objects.

Newtonian physics clearly fit well with mechanical philosophy; if anything, it brought up the prospect of mechanism gone mad. A clockwork universe excluded occult influences, but it posed the question of how the divine will could manifest in the world. Newton himself felt the force of this question. He argued, for example, that while his universal law of gravity explained the orbits of the planets, the stability of their arrangement was due to divine action. God had to intervene subtly, and occasionally make sure the order of the heavens was maintained. Newton's pursuit of alchemy was also a search for a kind of spiritual action that made sense in a world of physical laws. Early scientists had no natural explanation for complex order, and they often invoked mysterious "spirits" to explain what we today understand as chemistry. Diderot suggested that gravity showed that matter arranged itself in an orderly fashion through natural laws, but this was far from providing a satisfactory explanation. Before the time of Darwin, scientists often found themselves pointing to complex order, especially that seen in living things, as a product of supernatural intelligence. Complexity was the work of God, either by special creation or *through* natural law.

NONBELIEF COMES OF AGE

In the nineteenth century, in the industrialized West, nonbelief approached its current form. Skeptical philosophers continued chipping away at both conventional theistic metaphysics and the newer ways their devout colleagues devised to defend the faith. Many philosophers moved on to other matters, particularly worries about the cultural crisis made manifest by the slow erosion of religious belief in Western Europe. Moral objections to religion found a larger audience; secularism developed a real constituency. And in all this, the perception that the new science had undermined traditional religion took on a central role.

In an increasingly industrialized, urban world, established religious institutions faced difficulty reproducing traditional beliefs, especially since orthodox religion was closely identified with the interests of the privileged classes. So secularist ideas could develop a popular appeal. Left-wing thinkers presented scathing moral and political criticisms of religion, often embracing materialist views for what appeared to be their liberatory potential. Even some in the middle classes, predisposed to favor religion as a way to keep social order, could be interested in hearing critiques of God. The United States, noticeably more devout than Europe even in the nineteenth century, never experienced the same level of working-class disillusionment with religion, and its government suppressed any socialist tendencies recent European immigrants attempted to transplant to the New World. Still, many respectable Americans of the nineteenth century enjoyed the speeches of Robert Green Ingersoll, the "Great Agnostic." A famous orator and politician—and certainly no socialist—Ingersoll skewered popular religious beliefs in a way no public figure can get away with today. And whether on the political left or right, secularists equated science with rationality sweeping away the superstitions of yesteryear.

Intellectuals fell away more rapidly; even if it is hard to say that most distanced themselves from religion, enough did to make the crisis of faith a common theme in nineteenth-century intellectual life. Literary figures explored nonreligious options to secure morality and meaning in a changed world. Secular substitutes for religion flourished in the nineteenth and twentieth centuries, promoted by thinkers enamored of anything from messianic socialism to market worship. Scientists often remained more conservative. Even so, modern moral philosophers could start their work by assuming that science had made the traditional religious vision no longer believable.

Although science helped weaken traditional, scriptural, authority-based religion, the primary reasons for the decline of religion in Western Europe have little to do with science. Far more important were the bureaucratic rationalization of the state and economic production, the fragmentation of social life, and political changes (Bruce 1996). Nevertheless, Western religious institutions have also depended on elaborate doctrines—intellectual schemes—to help maintain their influence. More than weightless metaphysical speculation about transcendent realms, these doctrines have been anchored to concrete claims. Jews, Christians, and Muslims have taken Adam and Eve, Noah's flood and so forth to be historical. They have understood answered prayers, angels and demons, and the immortality of the soul as definite realities, not just moral metaphors. These were important beliefs, and by the nineteenth century it had become obvious that these beliefs were seriously challenged by modern science. Science, moreover, could not be easily ignored, as tangible results of the new knowledge in the form of new technologies were penetrating everyone's lives.

Many devout thinkers responded by making their religious claims more concrete and literal, trying to present them as facts as solid as anything science had discovered. This empirical-mindedness would lead both to an increasingly brittle fundamentalism and to a deepening nonbelief if fundamentalist apologetics no longer appeared convincing (Turner 1985). Alternatively, more liberal theologians decided to bypass science, reinterpreting traditional doctrines and emphasizing the flexibility of religious metaphors. Some, such as Friedrich Schleiermacher, argued that the foundations of religion were in emotion; Rudolf Otto (1923) became enthusiastic about nonrational (but not, he insisted, *irrational*) religious responses to mystical experiences. More liberal ways to reconcile science and religion, however, could also lead to nonbelief, if all the reinterpretation left a watered-down faith with little content worth defending. In either confronting science or avoiding it, serious religious thought began to contend with nonbelief as a live option.

The intellectual challenge science presented was now manyfold. First of all, physics weighed in, as the most mature and most basic science. Newtonian physics reached its peak in the nineteenth century, explaining an impressive range of natural phenomena and leading many to think it must capture the fundamental nature of the physical world. Though physicists often remained personally devout, there was no longer any need to refer to any gods while doing physics. Newton had once thought the stability of planetary orbits required regular divine adjustments; Pierre-Simon Laplace demonstrated this long-term stability entirely within physics. The story goes that he presented his book *Celestial Mechanics* to Napoleon, who asked what the role of God was in his theories. Laplace is said to have replied, "I have no need of that hypothesis." That the alleged creator and sustainer of material order should become an unnecessary hypothesis in physics was bound to bolster nonbelief.

Religious doubt also lurked in the grander Laplacian vision of a closed universe, where the natural causes of physics were sufficient to explain all—where the impersonal order of material objects interacting with one another determined everything. A nineteenth-century physicist defending God could still embrace the deistic option, saying God had created the clockwork universe, wound it up, and left it to run on its own. Deism, however, was religiously unsatisfactory, and its influence had waned considerably. The devout physicist was more likely attracted to the tradition of natural theology, which also reached its peak in the early nineteenth century. Theologians and scientists would join forces describing how divine providence was obvious in the design of the world, claiming, with William Paley, that just as it is obvious a watch found on a beach is a product of a mind, so must the infinitely more wondrous and complex universe be a result of intelligent design (Paley 1802; on natural theology see Olson 2004). On top of this evidence, there was the solid historical record of revelation recorded in the scriptures. Still, physics on its own no longer contributed to the feeling that there must be a God, and materialists could use physics to defend their suspicion that natural causation sufficed on its own.

Trust in religious historical narratives also became difficult to defend. Critical historians taking a self-consciously scientific approach had been examining the scriptures since the Enlightenment; in the nineteenth century, their skeptical conclusions broke into the academic mainstream. In this, they were aided by scientists who had been trying to understand the history of our planet. Geologists discovered "deep time": that the earth had a much longer past than religious tradition had imagined. Mountains had formed and eroded away, species very different from those living today had thrived and gone extinct. It became strange to think that Adam had named all the animals at the dawn of time just six thousand years ago, or that Noah's flood had covered all the earth even more recently.

And then, Darwin. Though Newtonian physics had already led toward making God an unnecessary hypothesis, Darwinian evolution was historically the greater challenge to religion. When it became clear that species had originated and diversified by descent with modification, no remotely literal reading of the traditional creation stories could remain intellectually respectable. The story of sin and salvation, the unique place of humans in creation—many doctrines vital to Christianity—had to go back to the drawing board. And science finally had something weighty to say about how complex, intricate entities such as living things came about. No special creation was necessary; in fact, evolutionary biology explicitly avoided bringing any intelligent designer into its explanation of the diversity of life. Natural theology was in trouble.

Evolutionary theory immediately caused religious turmoil. More traditional religious views came under threat, and since they were widely popular and represented the orthodoxy of the time, there was a significant uproar (see Figure 1.3). But even among evangelical Christians, many theologians found ways to accommodate evolution. They interpreted evolution as an inherently progressive, divinely

Figure 1.3 Charles Darwin caricatured as a monkey hanging from the tree of science, in response to the publication of his *The Descent of Man*. Darwin's theory of evolution had an enormous cultural impact in the late nineteenth century. (André Gill, cover of *La Petite Lune*, no. 10, Paris 1871. Library of Congress Rare Book and Special Collections Divisions)

guided process leading up to creatures with spiritual qualities (Livingstone 1987). In fact, the starkly naturalistic, Darwinian version of evolution did not catch on at first. Most biologists, though accepting descent with modification, were not convinced that natural selection could be truly creative. Many saw evolution as the progressive unfolding of a structural order inherent in nature (Bowler 1988). Such views retained a divine influence on nature, though this divinity was more like the remote metaphysical force of philosophers than the special creator of religious tradition. Biology would not join the physical sciences until the 1930s, when purposive forces driving evolution dropped out of play.

While natural scientists were moving toward explaining the world without divine powers, social scientists also contributed to the growing intellectual stature of nonbelief. One of the fondest hopes of Enlightenment thinkers was to use science to liberate people from custom and tradition, to order social life on the basis of true knowledge rather than superstition. The social sciences, which took shape after the natural sciences, grew out of this Enlightenment hope, and often had a secularist thrust. Enlightenment social thought such as the economic philosophy of Adam Smith moved beyond traditionally moralistic and theological frames of reference. Indeed, secular social thinkers tried to explain religion in social terms rather than as a direct response to supernatural realities. Some of the best-known nonbelievers of the nineteenth and early twentieth centuries, such as Ludwig Feuerbach, Karl Marx, and Sigmund Freud, tried to identify the psychological and social factors they thought sufficed to generate and sustain religion. Sociologists such as Emile Durkheim treated religions as human phenomena, asking what their function was in our societies.

The social scientists did not have to dismiss the divine. Many remained uncommitted, stating that examining the social and psychological causes that shape religion did not mean denying the supernatural claims those religions make. Bringing religion into the natural world, however, could not help but make it look less miraculous. And many social thinkers went further, portraying religion as a primitive stage in a progressive evolution of humankind, or as a projection of human needs or anxieties into the universe. Secularists began to express confidence that religion was on the wane. If not dead yet, God was surely dying.

As a result, the end of the nineteenth and the beginning of the twentieth centuries could boast a well-developed scientific materialism. For example, Ludwig Büchner's *Force and Matter* reached its fifteenth edition in 1884 and was a very influential and comprehensive statement of this mature materialism. Büchner and his fellow materialists were very impressed by nineteenth-century science, and although they admitted their vision was not purely scientific, it was clearly inspired by science. They confidently drew on the physical sciences to express their view of a world composed of material objects obeying natural laws, embraced evolutionary biology, and attempted to extend this picture to human experience, the last bastion of mystical views.

Nineteenth-century materialists were especially concerned to show that consciousness was a product of material processes, although they were not wholly successful in their efforts. Their approach, in keeping with the science of their time, was more biochemical. They attacked vitalism, the notion that life is animated by a mysterious force, denying the existence of any life force. And since physics and chemistry were beginning to explain the previously mysterious processes of life, nineteenth-century materialists had hope that the mind could also be understood as a natural phenomenon. In hindsight, many of their proposals look strained today, but the overall thrust of the materialists was more important than any of their particular psychological claims. They could not promise an immediate solution to all puzzles, but they insisted the mind was not a separate principle or an impenetrable mystery. They thought that with science, we had a good prospect of learning more. Materialists could not claim there were no gaps in our knowledge; instead, they said we could get somewhere in closing these gaps. Science made steady progress in figuring out what used to be unknown, and materialists said that this progress was a good indication that we live in a world where everything is the result of natural forces inherent in matter. Much has changed since Büchner's time, from the details of our scientific theories to the weakening of the sharp separation he assumed between philosophy and empirical knowledge. But his attitude toward science remains central to science-inspired nonbelief today.

If not a golden age of nonbelief, the nineteenth and early twentieth centuries were still good to infidels. For the first time, doubters found themselves in an environment where outright skepticism had spread throughout intellectual circles, and many ordinary people had also rejected religion. More than at any time before or since, this was an optimistic time for nonbelief.

Even so, there was no triumph of science over religion, and even where organized religion had declined, no mass abandonment of the supernatural. Though science was widely acclaimed, it did not impress everyone. Scientists thought their work revealed the secrets of an awesome universe, but others thought they "unweaved the rainbow" and deadened a once sacred world (Dawkins 1998). Technology helped people fly, but also created the poet William Blake's "dark Satanic mills" and industrial wastelands. For nonbelievers, science was typically an icon of reason and a tool to achieve heaven on Earth; they capitalized Scientific Progress without irony. Not all agreed, and not just because of attachments to old-time religion.

Furthermore, spiritually minded thinkers also saw opportunity in the new science and the way the old orthodoxies had become implausible. The occult, individualist alternative to both traditional religion and scientific rationalism never vanished, and it regularly surfaced as a third option. Just as the occult ideas of Mesmerism had become the rage in late Enlightenment times, merging the promise of science and natural magic, psychical research captured the imaginations of both the public and some scientists in the late-nineteenth century. Materialists promised freedom but denied the spirit; those who found the promise hollow but remained impressed with science tried to harness science to probe spiritual realms. Psychic phenomena always hovered on the fringes of respectability, but its attraction was undeniable. For example, Annie Besant, fiery atheist pamphleteer in her youth, ended up heading the Theosophical Society, a leading occult organization. If the advancement of science created social openings for nonbelief, it also encouraged occult attempts to harmonize science and spirit.

Orthodox religion slipped somewhat in intellectual respectability, but it survived, even thrived. Even the old scientific-style arguments for God never vanished; they just reappeared in new forms. Nonbelievers may have thought Hume savaged the ancient argument from design, Newtonian physicists made it irrelevant, and Darwin finally buried it. But in the 1930s, renowned astronomer Sir Arthur Eddington could still argue that the second law of thermodynamics showed the universe had a creator. Since the universe was becoming more disorderly over time, this meant that everything was degenerating, winding down from a more perfect state that could not have been achieved by strictly physical means.

Modern times meant nonbelief acquiring some respectability. But for science, the main consequence was a new balance of power. Conflict between scientific and religious institutions could occasionally flare up, but the two largely managed to negotiate separate spheres of influence. Scientists claimed the natural world, liberal theologians the world of morality and meaning—the bigger picture behind all the mere facts the scientists produced. As a result, the conventional wisdom about the relationship between science and religion erected a wall of separation between the two. Provided each stuck to their proper domains, science and religion could not conflict, or even say much of use to one another. As long as scientists were careful not to go beyond their data in making claims about spiritual matters, and theologians kept quiet about how the material world functioned, everyone could live happily ever after. Nonbelievers had hoped for science triumphing over religion; they had to make do with a cold peace instead.

SCIENCE AND NONBELIEF TODAY

Nonbelievers hoped that religion would fade away. This has not happened. Today is a time of extraordinary religious vitality. Many spiritual traditions seek converts all over the globe, reinventing themselves in doing so. The monotheistic faiths especially are doing well. Though Christianity has become moribund in the Mediterranean basin and in Europe, it remains strong in the Americas and is growing tremendously in Africa and Asia. Countries with a majority Muslim population, where once a slow secularization was thought to have set in, have been re-Islamized with great fervor. Other faiths have responded, in forms like the Hindu fundamentalism which is undermining the secular political tradition in India.

In all this religious activity, the most conservative, miracle-proclaiming, fundamentalist religions have been among the most successful. Liberal theologians had worried that their faiths could not survive unless they reached an accommodation with science and downplayed their supernatural elements. Today, the religions whose doctrines seem scientifically most absurd are not only doing well, they are doing so by exploiting the most modern technologies.

Sociologists of religion hotly debate what all this means; some ask whether there has been much erosion of religiosity even in the most developed countries. Only one large region, Western Europe, seems to have clearly become more secular, along with Japan in Asia. Religion is no longer significant in public life in much of Europe, where vast numbers of ordinary people do not participate in religious life

any more and churches are being converted into night clubs and re-
tail outlets. Surveys indicate that substantial segments of the population disbelieve in any personal God; numbers approaching about 50%
in countries such as France and the Netherlands are typical (Norris
and Inglehart 2004). About 10 to 20% disavow any supernatural belief: not accepting even a vague life after death or some unspecified
and distant creator-principle behind the universe. This is not to say
Western Europeans have turned into scientific rationalists; it appears
that many have just dropped out of religion. The gods have not so
much been chased out of a universe of force and matter as become irrelevant to daily life. Though occult, individualist spiritual beliefs
have grown in influence, they have not been able to conquer the social territory vacated by organized religion.

The United States is, in contrast, strongly religious. Religious participation remains high, supernatural beliefs influence public policy,
and polls regularly show about 90% of Americans expressing belief in
a very conventionally conceived God. American religion is more fragmented today, and a growing minority are dissatisfied with organized
religion even though they continue to hold strong supernatural beliefs. In 2004, The Pew Forum on Religion and Public Life found 16%
of respondents identifying with no particular religion. Many people
today consider themselves "spiritual but not religious," and often are
attracted to the disorganized supernaturalism of New Age and occult
beliefs. Completely godless individuals, however, do not add up to
more than about 6% of the population (Norris and Inglehart 2004).
Though many public institutions have been secularized, and an em-
battled tradition of church-state separation is still standing, the United
States remains a culturally Christian nation.

The worldwide vitality of religion coincides with a severe decline
in the fortunes of secular philosophies of liberation, particularly socialism. Communism as it developed in twentieth century totalitarian
states may have been only a travesty of socialist ideals, but its collapse
nevertheless triggered a crisis in left-wing political movements in general. Today, explicitly secularist social thought, long associated with
the political left, has trouble attracting a constituency beyond those already comfortable with a secular way of life. In fact, the standpoint
of religious faith, looking beyond worldly advantage, seems one of
the few still available today to resist the more rapacious aspects of
capitalism.

The political failure of the most ambitious forms of secularism has
also affected intellectual life. Among many social thinkers, Marx and

Freud were once treated as oracles on the nature of all things, and their negative pronouncements on spiritual beliefs carried a lot of weight. Now, they are seen to have failed. Science contributed to this failure. In economics and sociology, Marx had some interesting insights, but much of what he proposed appears plain wrong. As for the Marxists, they were always more prophets of worldly salvation than scientists; "scientific socialism" was neither. Freudian ideas were also popular once, but experimental psychologists always kept their distance. Eventually, it became clear that psychoanalysis was severely flawed; in many ways it did not deserve to be called a scientific idea at all. Once again, nonbelief was linked to a secular ideology that acted as a crude substitute for religion.

Though scientific institutions remained independent of political forms of nonbelief, today science also suffers from a tarnished image. The days of speaking of scientific progress as an unalloyed benefit are past. If once we were sold on "better living through chemistry," we now have become more aware of better killing through nuclear physics. Scientists have always promoted their work by pointing to the technological benefits to which new knowledge led. In practice, modern science became a large-scale undertaking supported by its services to commercial and military enterprises. It would be amazing if scientific institutions wholly escaped criticism directed toward the incessant war making, environment despoiling, and shallow money-centeredness of modern civilization.

Enlightenment philosophers held up secular government as a way to liberate us from priests, natural science as a way to obtain the knowledge to improve our lives, and secular social science to help us live together. The driving force behind nonbelief has very often been an ethical objection to religion, and only secondarily the suspicion that the world was not in fact shaped by spiritual powers. Today, many look back and perceive secularism turning into totalitarian ideology, natural scientists devoted to finding ways to blow ourselves up, and social science developing ways to manipulate us through advertising and worse. Though it is an exaggeration to lay all this at the feet of nonbelief or science, the ethical motivation for religious skepticism or for championing science has diminished in intellectual circles.

The faltering of secularism outside of Europe has also given new impetus to attempts to better harmonize science and religion. If science and religion could support one another—going beyond the cold peace of "separate spheres"—they both might benefit. Restoring soul to science would help science recover the moral high ground and allow religion to develop more intellectual substance than either fundamentalism or the kind of theology that disengages from science entirely. With the financial support of new institutions like the John Templeton Foundation, the search for harmony between science and religion continues to attract many thinkers.

The question of whether science and religion conflict has interested social scientists as well. In this age of the opinion survey, one way to settle the issue might be to ask scientists themselves. They, after all, should be in the best position to know. The data, however, turn out to be ambiguous at best and hard to interpret.

For example, sociologists Rodney Stark and Roger Finke point to a 1969 survey of 60,000 American college professors which shows that "a very substantial share of faculty engaged in doing science . . . are quite comfortable with religious faiths" (2000, 55). This and other very broad surveys of natural-science-related professionals, including engineers and industrial scientists, find that while significantly more indicate "none" as their religious preference than the general U.S. population, they are only a minority at around 30%. Sociologists like Stark and Finke argue that such data demonstrate not only a social accommodation between scientific and religious institutions, but the absence of any intellectual conflict between modern science and supernatural belief.

The picture, however, is not so clear as that. Surveys more narrowly focused on elite scientists, such as one done on members of the U.S. National Academy of Sciences, find "almost total" rejection of spiritual beliefs such as a personal God or human immortality: in the NAS survey, less than 10% expressed a spiritual belief (Larson and Witham 1998). It appears that among scientists who are leaders in their fields, nonbelief is overwhelming. While an industrial chemist or a biology teacher in a religiously affiliated college can easily have religious beliefs that are close to the wider community, the elite scientific culture is very secular. The popular literature of nonbelief, naturally, prefers to dwell on such data, interpreting it to mean that the conflict between science and religion is very much alive.

Such surveys, unfortunately, say very little about any *intellectual* friction between science and religion. Certainly, the science departments of modern universities are very comfortable places for nonbelievers, as are philosophy departments. Nonreligious scientists can safely assume many colleagues share their skeptical outlook, and a generally secular ethos prevails in any case. On the other hand, sci-

entists are professionally rewarded for competence in narrow specialties. For example, a physicist who is brilliant in the lab might also
be a conservative Mormon and publicly argue that early North American history took place in the way orthodox Mormons believe. His
views would be considered crankish among academic historians and
archaeologists, but among physicists, they would remain irrelevant as
long as they did not interfere with his work in physics. A poll of scientists does not say much about science and religion because surveys
cannot address whether scientists have well-thought-out reasons for
either their disbelief or devoutness.

So today, the connections between nonbelief and science remain
murky. Political forms of nonbelief have lost steam, but although this
means there is now less of a popular constituency for religious skepticism, science is done within an elite culture that does not directly
reflect prevailing social attitudes. This scientific culture can accommodate religion or keep it at arm's length, but even in an intellectual
environment like that of science, prevailing attitudes need not reflect
careful arguments.

Discussions of religion are full of argument, and both the popular
and academic literatures today include many works on science and
religion. Naturally, the largest market share belongs to publications
supporting some form of harmony between science and the local religion. There is, however, skeptical material as well, and some influential scientists contribute to these, such as the biologist Richard
Dawkins and the physicist Steven Weinberg.

Skeptics today continue to argue along some historically very familiar lines. For example, the theme of reason versus revelation
emerges: nonbelievers often say that the critical approach inherent in
science does not allow an attitude of faith, in which certain claims
about the world are held true without evidence and without the possibility of revision. Religion, they argue, rests on faith, and hence does
not mix well with science. Another common theme is that religious
belief interferes with the progress of science. Religions are committed
to supernatural beliefs such as life or the mind requiring a nonmaterial spark. Such convictions get in the way of investigating things.
Moreover, religious influence can actively discourage certain areas of
science, as evidenced in political controversies over matters such as
genetic research today.

Though popular, such arguments are not entirely convincing, even
to audiences inclined to see science in a positive light. There are many

different religions, and those with more than a few followers generally accommodate many ways of being religious. Not every believer understands "faith" to be a blind commitment; they can take faith to be trust in a spiritual power experienced personally through a religious way of life. In any case, there has never been a lack of pious scientists who investigate the world as an act of devotion, motivated to learn more about the creator by studying the wonders of the creation. The diversity of religions also means that there is some sect likely to support almost any type of research for some religious purpose. If, for example, conservative Christians oppose human cloning, the Raëlian new religious movement will step in to try and do cloning themselves. "Religion" is not a monolithic faith attitude that invariably opposes scientific enterprises.

There is, however, a stronger argument nonbelievers put forth about science and religion, which is also more straightforward. That is the claim that through modern science we have learned a lot about how the world works, and that what we have learned indicates that this is a purely natural world, with no gods, demons, or spirits of any sort.

SCIENTIFIC NATURALISM

Science and nonbelief come closest together in the perspective of *scientific naturalism*, the claim that the world can be fully accounted for in terms of entities similar to those the natural sciences presently acknowledge. At a minimum this means no God, no ghosts, no specially spiritual realities of any sort beyond the natural world.

Religions claim the existence of supernatural agents, whether these are ancestral spirits, angels, or gods. The so-called "world religions" take a more sweeping view in their official doctrines; they say that an all-powerful supernatural agent governs the world, or that in some perhaps hard to articulate way, the ultimate nature of reality is spiritual. Their theologies, abstracted from the much messier world of everyday religiosity, tend to portray the world in top-down, hierarchical terms. Mere matter, on its own, would be an inert lump of substance. Spirit is a separate and higher principle, which acts from above on matter to give it shape, to inject meaning and order into what would otherwise be chaos or nonexistence.

Naturalists invert this picture, taking a bottom-up view (see Figure 1.4). Down deep, naturalists say, the universe is made up of the sort of objects that scientists, particularly physicists, deal with. Furthermore, physicists describe the interactions of these objects in terms of combinations of random behavior and lawful patterns, without any purpose or personality. Naturalists think that the complex, orderly, and interesting aspects of our world, including life and mind, are unthinkingly assembled out of this physical substrate.[2] As philosopher Daniel C. Dennett puts it, naturalists obtain complexity by building "cranes" from the ground up, not by seeking "skyhooks" descending from above (1995). And so they expect to find just chance and necessity at the bottom of everything, including the intelligent actions of personal agents such as ourselves.

This view is clearly inspired by the success of the natural sciences. We have found that fundamental physics gives rise to many complex phenomena. This includes setting the stage for chemistry. The rules chemists use in their work are derived from physics; we need invoke

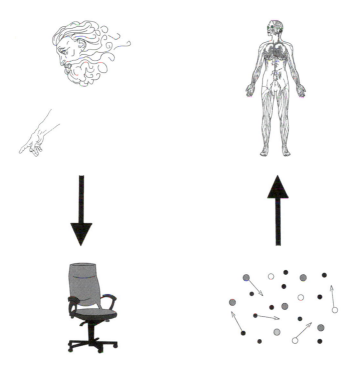

Figure 1.4 In a top-down picture of the world, a spiritual force acts from above to create or give shape to otherwise inert matter. A chair, for example, is obviously designed for a purpose. In a bottom-up view, even humans are entirely made up of interacting physical objects. (Ricochet Productions)

no "chemical souls" to understand molecules. Biology, in turn, can be based squarely on the physical sciences. The processes of life are entirely physical, requiring no "animal soul" or "life force." Ever since Darwin, we have also understood how biological complexity arises naturally. Naturalists think the prospects are very good to extend this picture across the board, explaining minds, human societies, and indeed everything in a way that is continuous with the natural sciences.

This does not mean physics is *the* basic science and the rest are stamp collecting—merely collecting and organizing isolated facts. Different sciences test each other and their results feed back into one another. And some of the most interesting ideas in modern science concern how to assemble complexity: theories about order and disorder, information, or evolution are not restricted to any one discipline. Naturalists, therefore, tend to see all areas of inquiry as being connected to one another. Knowledge does not proceed from out of physics any more than out of metaphysical first principles handed down by prophets or by philosophers. Dennett describes his approach as "*naturalism*, the idea that philosophical investigations are not superior to, or prior to, investigations in the natural sciences, but in partnership with these truth-seeking enterprises, and that the proper job for philosophers here is to clarify and unify the often warring perspectives into a single vision of the universe" (2003, 15).

A promise to find unity in knowledge is very attractive, and this is a reason many scientists and philosophers are drawn to a naturalistic perspective. More religious approaches may also promise unity—such as all knowledge coming under the umbrella of Christian theology, the "queen of the sciences" in medieval times—but naturalism, many judge, delivers more. All of our sciences, including disciplines such as history, have adopted naturalistic perspectives in the last few centuries. The serious theories in play today no longer involve cosmic purposes or supernatural agents. This has coincided with an explosive growth in knowledge. Even scientists who are personally religious work in a naturalistic context within their own disciplines, and take naturalistic theories to have the best prospect for further progress. It is no great strain to think that the success of naturalism in all our sciences is due to our world being a bottom-up, godless, and ghostless place.

Conservative religious thinkers are tempted to challenge the widespread naturalism within science by denying that this success is all it is advertised to be. They claim, for example, that many areas of science are stagnating because of overcommitment to naturalistic ideas like Darwinian evolution. They believe science can be improved by bringing "intelligent design" back to biology and other sciences.

Theological liberals usually prefer to avoid conflict with science. They often respond to naturalism by saying science is naturalistic by definition. Science is concerned with seeking natural explanations for natural phenomena. Hence, science is methodologically naturalistic: the supernatural does not properly come within the scientific domain. The sciences therefore have nothing to say one way or the other about the truth or falsity of claims like the existence of a God. Questions concerning ultimate meanings, gods, and spiritual realms are metaphysical and not within the competence of science.

Such a view allows an easy peace between scientific and religious institutions and so is very often the conventional wisdom. Unfortunately, it has problems. Religious people make many claims about alleged spiritual realities that can be investigated scientifically. They may favor special creation over evolution, claim that psychics or saints display supernatural powers, or say that the physical universe shows signs of its creator. Surely the success or failure of such claims must have some bearing on whether their ideas about spiritual realities have any substance. The only way to avoid this would be to present a bloodless ultraliberal metaphysics as the core of religion, which is implausible. It also flirts with making religious claims meaningless, as the claimed ultimate realities then would make no difference in the world as we experience it.

Indeed, any attempt to limit the scope of science and thus protect the gods from criticism must remain dubious. The main reason is that a God untouchable by science becomes a cosmic Santa Claus. Once we figure out what is really happening, we understand Christmas within the everyday world of traditions, culturally significant stories, and parents placing glossily wrapped presents under gaily decorated trees. An actual Santa Claus associated with elves, reindeer, and a workshop at the North Pole no longer plays any role in our explanation of Christmas; we say the jolly old fellow is not real. A sophisticated Clausian could try to say that Santa Claus remains real, that he is still the spiritual driving force behind Christmas who acts *through* the parents placing gifts under trees. This view, however, would add nothing to our understanding of Christmas, and unless we are emotionally very attached to the belief in Santa Claus, we are likely to consider this sophisticated spirit of Christmas not worth our time.

Religious beliefs disconnected from scientific reality checks are asking for the same fate as Santa Claus.

Scientific naturalism, then, presents a serious intellectual challenge to religion. It cannot be waved away with philosophical arguments: naturalism's challenge is plausible or not depending on the force of the specific scientific arguments naturalists bring together to try and make their case. This means examining modern physics and cosmology to see if our universe requires a creator or a spiritual underpinning. We have to look at biology and ask whether Darwinian evolution chases God out of the world and accounts for creativity without anything from outside of nature. Cognitive science and artificial intelligence research will come into play, telling us what the prospects are for understanding minds without any nonphysical souls or spirits. We will have to see if miracles or psychic phenomena really happen. And we will have to venture into the social sciences to find out how naturalistic approaches to explaining religion itself work out.

NOTES

1. Greeks get more emphasis in histories of science and nonbelief, even though ancient Chinese and Indian thought also includes intriguing proto-scientific ideas and hints of materialism and nontheistic philosophy (an exception is Hecht 2003). Buddhists, for example, developed a very sophisticated introspective psychology in a nontheistic context. However, modern nonbelief, especially science-inspired nonbelief, has predominantly Western roots.

2. Strictly speaking, this is a *physicalist* (Melnyk 2003) view, rather than a "non-reductive naturalism," which might, for example, allow for minds that are not explainable in any physical terms yet are fully natural. A full-blown physicalist approach, however, is both more in tune with today's science and with the kind of nonbelief that does not countenance any spiritual entities. The kind of naturalism that attracts most scientists (rather than philosophers) is more often the physicalist variety.

Chapter 2

An Accidental World

WHY PHYSICS IS HARD

As generations of students rediscover all the time, physics is difficult. The mathematics is a bother, and all too often, the subject appears as an endless series of formulas and abstract problem-solving techniques remote from everyday experience. For too many, the dreaded college physics course seems designed to produce mediocre grades even after diligent study and application.

The reason for this is that the human brain is not built for doing physics. It is not that physics is exceptionally complicated; if anything, physics is remarkably simple and compact in its mathematical structure. Though its formulas at first look like a disconnected list of equations applicable to many separate circumstances, the overarching conceptual schemes that lie at the heart of physics richly connect everything together in an elegant overall picture. Physics, however, cannot easily exploit the natural ways our brains grasp the world; its conceptual schemes remain thoroughly abstract and alien. After some arduous undergraduate training, those students who avoid being turned off only begin to appreciate the overall structure of physics. Others might read popular science books or take a "physics for poets" course exploring the concepts of physics with a minimum of math. This can lead to a genuine appreciation of physics. Nevertheless, at this level, a physics enthusiast will still have difficulty distinguishing between the strange ideas in modern physics and a scientifically asinine plot device for *Star Trek*. A black hole seems just as bizarre as a transporter beam, though one is real and the other a physical near-impossibility.

We all know enough about how objects move or collide or fall to get around in everyday life. Cognitive science envisions our brains as having many faculties or modules common to our species (Pinker 2002; Stanovich 2004, chapter 2), including one for "folk mechanics," which generates expectations about the movement of inanimate objects. These expectations are usually not articulated: when we catch a ball we just do it, without conscious calculation. When people do express their commonsense views about mechanics, we get notions like movement requiring a continual force or impulse. And folk mechanics is applied through training in specific skills, not by learning equations. Humans can become very good at throwing and catching objects and predicting their trajectories. But learning how to explicitly calculate projectile motion comes hard to us. With simplifying assumptions such as ignoring air resistance, the math becomes easier, but learning to do the kind of abstraction by which a physicist simplifies a problem is still difficult.

The equations are vital to physics as a science, as they allow us to generalize abstract representations to find the common patterns in the physical world. Folk mechanics, though wonderful at letting us make effortless, rapid decisions in situations relevant to our evolutionary history, does not generalize. We could never notice the commonalities between the motion of planets and rocks if we were limited to repeatedly training the neural networks in our brains to perform sets of skills based on our innate folk-mechanical capabilities.

So learning physics involves a lot of unlearning as well. Students encountering Newtonian mechanics have to understand that even though their folk-mechanical intuitions scream otherwise, if no force acts on an object, it continues to move at a constant speed in a straight line. When we take friction and other complicating forces into account, the same principles apply both to planets and to dragging a table across the floor. More advanced students repeatedly find out how the world is fundamentally *not* as folk mechanics leads us to expect. Beyond the domain of everyday life, our intuitive physics fails. It is useless for understanding outer space, very low temperatures, very small particles, and so on.

Earlier forms of physics, before the scientific revolution, did not depart far from folk mechanics. Aristotelian physics, which was the best on hand from Greek times through the Middle Ages, held that objects came to a halt when the impulse imparted to them ran out. The stars had to have souls: their motion did not slow down, so they had to possess spirits similar to the spirits animating living things. On Earth, Aristotelians distinguished between horizontal and vertical motion, explaining why certain bodies rose or fell or sank according to the mixtures of air, fire, earth, and water that made up sublunar objects. Overall, everything moved in such a way as to seek its natural place. In fact, Aristotelian physics envisioned a physical universe infused with purpose. This was a universe ruled by causality, but Aristotelians looked beyond the mechanistic notion of cause and effect most familiar to modern people. A complete explanation also required knowledge of the "final cause," which was the purpose or function of an object or event (Grant 2004, 45).

Another module of human cognition is folk psychology. We readily identify personal agents and understand them in terms of their beliefs, intents, and purposes. In its way, Aristotelian physics attempted to unify folk-mechanical and folk-psychological modes of explanation; in principle, there were purposes behind everything. So it was natural to see the universe as the design of supernatural agents.

Compared with physics after the scientific revolution, however, Aristotelian physics did not offer much in the way of solid explanations or predictions. Being close to folk mechanics, it was easy to understand; even today, physics students find it easier to think in Aristotelian terms. Its vague descriptions of motion, however, were more of a loose framework for a grab bag of vague intuitions, skills, or more precise descriptions that did not generalize beyond very restricted domains.

Newtonian physics is much less intuitively accessible, though its notions of space, time, causality, and motion are not too far removed from common sense. We can still visualize the Newtonian clockwork universe as a place where particles interact like billiard balls bouncing off each other, but this "classical physics" is difficult. It already requires mastery of a specialized way of thinking remote from our folk-mechanical intuitions. And Newtonian physics is much more ambitious than what came before: classical physicists began to think about bringing everything under their style of explanation. Theoretical physics since the scientific revolution is distinguished by its expansive ambitions: far from being a way to solve certain types of abstract problems, an expandable set of practical skills, or a list of formulas and facts, it is an attempt to describe the fundamental nature of our world.

Modern physics, building on the developments in physics in the early twentieth century—relativity and quantum mechanics—is even more ambitious and further removed from intuitive physical notions. Today's physicists contemplate a world full of random events, where space, time, and causality are very different than how we naively imagine them. Especially with quantum mechanics, we encounter a realm that cannot be visualized in intuitive terms at all. The quantum world is completely alien to our brains; we can only approach it with the aid of some heavy-duty mathematics.

The comprehensive naturalism that is the most important form of intellectual nonbelief today rests heavily on the counterintuitive but expansive vision urged by modern physics. To understand it better, we need to delve deeper into the nature of physical thought today, and not just those physical ideas relevant to religious questions.

THE NATURE OF MODERN PHYSICS

Physics violates common sense. Physicists begin their work by relying on common sense, but well-developed physical theories are not constrained by what we intuitively think is reasonable. For example, there are no solid objects, at least not as we normally conceive of them. When a book rests on a table, the book and the table both consist of atoms that are almost entirely empty space. The reason the book does not penetrate the table and fall to the floor is that the particles that make up the atoms are electrically charged. There are no "contact forces" at the level of atoms; the book stays put entirely because of electromagnetic interactions.

There are some physical concepts for which it is difficult even to find an everyday analogy. Subatomic particles such as electrons have a property called "spin." Though this has some similarity to the spin of an everyday object like a top, it is actually quite different. The only way to understand intrinsic spin is to drop the analogy of a top—the electron is not an extended, classical object that can literally spin—and work through the equations of relativistic quantum mechanics. There is no way to experiment with the specifically quantum nature of spin without some highly specialized equipment set up to probe nature in ways far removed from everyday experience. Nevertheless, spin is a very important concept in modern physics, and much of what we see in our world would not make sense without elementary particles possessing the property of spin.

Religious ideas are often counterintuitive in a different way. A spirit, for example, is supposed to be a personal agent without a body. This violates our expectations about persons. Still, where spiritual beings are concerned, we readily make a host of inferences based on folk psychology (Boyer 2001). Religious concepts are easily learned, even by children, and unless they are stretched beyond recognition by philosophical theologians, they remain close to commonsense ideas.

From the point of view of modern physics, however, common sense does not count for much. When asked to give a reason for their religious beliefs, most modern people say that God explains our world. They favor the classical design or cosmological arguments, saying that the existence of a world, especially an orderly world, requires a cause beyond nature (Shermer 2003). It is only common sense to think that all we see cannot be an accident, that we have to ask what came before and caused the big bang that started our universe. It seems intuitively obvious that the fact that we do not have a formless chaos but a lawful universe points to a supreme intelligence behind everything. Such reasoning, however, relies on our everyday notion of causality. Modern physics presents us with a world where, contrary to our intuitions, everyday cause and effect are not fundamental but emerge from a substrate of random events. It undermines the common intuitions about order and chaos that supports religious beliefs.

Physics is deeply mathematical. Physicists seek precise patterns in nature: not everyday rules of thumb, not quick-and-dirty inferences that usually work, not judgment calls. This is not to say that everyday, non-mathematical reasoning has no place in physics. A physicist setting up an experiment does not sit down and calculate everything; she also relies on an intuitive feel for equipment and phenomena developed through much experience. Her calculations often cut corners and produce approximations rather than follow the exact standards of proof demanded by mathematicians. Nevertheless, she seeks to test mathematically expressed claims in a way that leaves little wiggle room in interpreting the results.

There is a popular belief that physics is limited because it relies on mathematics. Not everything, the claim goes, is quantity; focusing on what can be numbered and measured leaves out the more qualitative aspects of life. Such sentiments miss the mark. Mathematics can capture fuzziness and complexity as well as numbers. More important, mathematics is crucial for extending our reasoning into areas for which the basic structure of our minds does not prepare us. If the deep structure of the world were similar to the personal, purposeful aspects of social reality our brains are so good at grasping, we would not need so much mathematics. In a mature science like physics, the infinitely extensible language of mathematics frees us of the limitations inherent in our intuitive views of the world.

To give shape to ideas that go against common sense, we need to approach them mathematically. For example, one reason it seems crazy to think ours is an accidental world is that our brains are notoriously bad at dealing with randomness. So to make progress, we need a rigorously defined concept of randomness. And to put that concept to work, we need the aid statistics provides to our ordinary reasoning.

Physics is cumulative. In the course of research, scientists revise their theories and reject some old ideas. But by and large, physics progresses by extending our knowledge into new domains. Newtonian physics gradually starts failing as objects come close to the speed of light, but it remains a very good approximation at everyday speeds. In learning physics, students keep examining basic questions about motion, structure, and interactions at increasing levels of depth and generality. And what makes physics compelling as one learns more are the connections between the different concepts, how everything hangs together.

For example, consider momentum. A beginning student will encounter the Newtonian definition of momentum as the product of mass and velocity, $\mathbf{p} = m\mathbf{v}$. Then they might see one of Newton's laws expressed as the net force equaling the rate of change of the momentum, $\Sigma\mathbf{F} = d\mathbf{p}/dt$. Early physicists tried to figure out what the "quantity of motion" might be, and found out that momentum worked best. The equation $\Sigma\mathbf{F} = d\mathbf{p}/dt$ says that a force is what changes the quantity of motion. By figuring out the forces on an object and adding them up, we can figure out exactly how its motion will change.

There is, however, a lot more. Momentum also is significant because it is conserved: if no net external force acts on a collection of objects, the total momentum of the collection will not change. The reason is profound, and has to do with fundamental symmetries of the universe. If the laws of physics are the same in all places, so that an identical experiment performed here and a million miles away will produce the same statistical distribution of results, in such a universe momentum will be conserved.

Such a fundamental concept is worth knowing more about, beyond the collision problems of introductory physics texts. So a continuing physics student will learn about generalized momenta in increasingly abstract formulations of classical physics, and about how the notion of momentum changes subtly in relativity, when particles approach light speed. She will encounter momentum as not a number but an abstract mathematical operator in quantum mechanics, seemingly far removed from the Newtonian $m\mathbf{v}$ but closely connected nonetheless. She will develop a rich concept of momentum, taking its meaning from its role in the sweeping conceptual schemes of physics, the connections between momentum and other physical concepts, and the defining examples and important experiments that frame what momentum is all about. The structure of physics, and the strong connections between its concepts, is what makes physics so intellectually and practically powerful. It is *not* a list of formulas.

Physics requires reality tests. Experiments keep physicists honest: they want to understand how the world actually works, not build pretty mathematical castles in the sky. Being open to being corrected by nature, however, does not mean accepting experience at face value. Some ancient Greeks encountered their gods in dreams, and the most straightforward way of understanding such experiences is that the gods did, in fact, visit people in dreams. If we take experience to be delivering "just the facts," we will be stuck with folk theories and commonsense interpretations. These are very good for making quick judgments in everyday life, but they are poor models for the universe. Mature sciences reach beyond what comes naturally to us due to the structure of human brains. And so, doing experiments in physics is a sophisticated, controlled enterprise closely intertwined with theoretical work. An experiment often tests a specific theoretical claim and draws on an extensive background in physical theory to make sense at all. Otherwise, the numbers on dials and graphs on computer screens produced by an experiment would just be a meaningless jumble signifying nothing.

In the early twentieth century, some experiments involving rare decays of subatomic particles seemed to violate momentum conservation. Wolfgang Pauli proposed that the best way to account for the missing momentum was to introduce a new particle, the neutrino, which was escaping undetected. Indeed, the neutrino was a very ghostly particle, interacting with ordinary matter so weakly that a typical neutrino could routinely pass through the earth as if it were not there. So the initial evidence for the neutrino was weak. However, physicists got to work and found that a particle like the neutrino helped explain other experiments, and that it could play an important role in theories of subatomic particles. Today, we can produce neutrinos in the laboratory and experiment with them. Using very rare interactions of neutrinos captured in vast underground tanks of water surrounded by electronic equipment, physicists can even detect neutrinos from sources like the sun or distant supernovas. Though ghostly, the neutrino is now as real to physicists as rocks and apples.

As neutrinos were better understood, they became more than a possible explanation for one experimental curiosity—they were integrated into the overall structure of physics. The concept of the neutrino developed many connections to the rest of physics, and so it became more and more difficult to challenge the existence of neutrinos. A successful challenge would have to rip apart the whole theoretical and experimental structure held together by neutrinos and replace it with something better.

This is exactly how scientific reality tests work. Experiments are not one-shot deals producing indisputable facts that confirm or demolish theories. In physics, reliable knowledge is that which is integrated into the overall structure of physics. If we have independent lines of evidence supporting a claim, it becomes more secure. If theoretical and experimental work, interacting with another, converge onto an explanation, the explanation becomes persuasive. And when a new physical concept is successfully connected to many others, the evidence in support of the others becomes relevant to judging the merit of the new idea. Many physical claims thus become coercive for physicists: it becomes almost impossible to deny them. No physicist today would dispute the existence of neutrinos, even though they do not touch directly on our everyday experience.

A common misconception is that on the one hand there are experimentally verified "scientific facts," especially if they find technological application, and theories on the other hand. "Theory," we too often think, is not far removed from speculation; it certainly does not have the solid trustworthiness of a good hard fact. But if science, especially physics, restricted itself to those facts that directly affect everyday life, it would be mere stamp collecting. In physics, theory and experiment have to work hand in hand; either can correct the other. Experiments may sink a theory, but a good theory may also show us something wrong with an experimental design. *Theory is important*.

THE BURDEN OF PROOF

This is all very well and good—physics demands a different way of thinking and interacting with the natural world than what works for everyday life. And doing science is different from religious devotion as well: acts of faith, revelations, and trying to get on the good side of supernatural forces are not common themes in scientific work. But none of this has to motivate nonbelief. The approaches that work well for investigating the material world and those appropriate for relating to the spiritual realm might be different. A physicist can both design elegant experiments and utter heartfelt prayers.

Nonbelievers have been much impressed with how modern physics has no use for the supernatural. They have been inspired by how science has been so successful in understanding nature, in contrast to explanations that call on gods and demons. None of this greatly challenges religion, however, if science is limited to providing natural explanations for natural phenomena. If, for example, we have a sophisticated understanding of atmospheric physics today, allowing us to construct detailed computer models to study climate, this could conceivably cause discomfort for a religion that attributes droughts or storms to angry gods. But most modern religions allow their gods to be subtler. In any case, even if the gods are no longer directly in charge of the skies, this success of science does not touch on the metaphysical and existential worries that drive much of religion. The greenhouse effect is fascinating to the specialists, but it does not tell us much about death, suffering, and questions like who made the world and why. Perhaps science is limited.

If physics were stamp collecting—gathering facts and making lists of practically useful equations—it would be easy to see its limits. Consider old-fashioned natural historians, who, well into the nineteenth century, were busy cataloging new species without much concern for placing them in the larger theoretical context of biological evolution. They could not contribute much toward a more comprehensive understanding of the world, other than observing that life was diverse and intricate, and that this included plenty that was cruel and nasty as well as beautiful. But since modern physics derives its power from making connections, from relating everything to the structures of its overarching conceptual schemes, it cannot avoid being considerably more ambitious than natural history. Atmospheric science is not just a collection of facts about climate and the weather. Atmospheric physicists get their basic equations from the overall structure of physics, and their work on the complicated dynamics of the atmosphere feeds back into more general investigations of the physics of complex systems. It is not possible to do physics *just* about the atmosphere.

Since physics has a tendency to continually expand its scope, to ask fundamental questions, and to wander far away from common sense, physics is naturally ambitious. It is hard to place limits on it. After the scientific revolution, sophisticated mechanical explanations of nature dampened enthusiasm for more occult approaches. Mechanistic thinking became so successful, however, that it also took over territory claimed by orthodox religion. Soon scientists were asking questions about the nature of life and the universe as a whole.

Even today, physicists have a way of getting involved not just with investigations into superconducting materials but also with questions concerning life, the mind, information, computation, cosmology, and just about everything. Indeed, singling out physics as special in this regard is misleading: if physics has any preeminence, it is mainly because it is the most mature of our sciences. All of modern science has become much more like physics, from its use of mathematics to the central role of the interplay between theory and experiment. All our natural sciences forge connections with one another; physicists borrow ideas from biologists or computer scientists as well as the other way around. And as this happens, the strongest naturalist view, that our world is fundamentally physical, becomes increasingly compelling.

Consider the common, often religiously central belief in souls. Our mental life, it usually seems, is beyond mere material nature—our very selves are supernatural. Now, even in commonsense terms, the supernatural is counterintuitive; it violates ingrained expectations about persons having bodies and so forth. However, this sort of rupture in the natural order is very easily integrated with common sense. Folk psychology makes it very easy for us to think mind is radically different, beyond the capabilities of matter. Common sense says that we are souls housed in bodies, and raises the possibility that souls can be detached from bodies.

From today's scientific point of view, however, belief in souls is most likely mistaken. The idea that consciousness does not require any special soul-stuff is not new, but until recent times, this was philosophical speculation by a handful of skeptics. It paled in comparison with the force of folk psychology. Today is different, and the implausibility of the soul has as much to do with physics as it does with progress in understanding the brain.

Postulating a soul means claiming a rupture in the way physicists describe the world. It is to say that physics works well up to a point, but that when it comes to our thoughts and feelings, something radically new comes into play. Furthermore, this unphysical thing is supposed to act on the material world. Modern physics pictures a fundamentally impersonal world, explaining everything in terms of combinations of chance and necessity. A soul would not be a minor exception but a disruption of the structure of physics, turning the bottom-up picture of naturalism upside down. And since the overall structure of physics is so well supported and connects so many different phenomena, an exception will not be admitted unless there is very good reason.

Imagine that someone claimed to have discovered an exotic metal that was not affected by gravity. This would not be a routine discovery, providing just another fact to add to a long list of scientific facts. The existence of such a metal would radically challenge the current structure of physical theory. It would be a revolutionary claim that would need some very strong evidence before it could be accepted. It could turn out to be true, but the claim would face a heavy burden of proof.

Souls and other spiritual realities also face a heavy burden of proof. This is not to say that the issue is settled; we are nowhere near a full-blown scientific explanation of minds. But, we have already learned a lot, and scientists have a good idea about what kind of research promises further progress. The notion of souls has no such prospects. There is also historical precedent in favor of a materialist research program concerning minds. The materialists of the nineteenth century were very interested in understanding life in physical terms, because at that time life looked like something mysterious that could not be captured by physics and chemistry. Until just about a century ago, the notion of a life force or "animal soul" was still taken seriously. Today, we have learned much more about biology and integrated it with the physical sciences, and no serious scientist speaks of a life force animating living things. Today's materialists about the mind expect much the same to take place with the soul that is supposed to give us the spark of consciousness.

Because of its successful track record, and its comprehensive ambitions, modern physics turns out to be central for the case for naturalism, and hence, nonbelief. It is very difficult to limit physics to a religiously harmless corner of reality. And if today's structure of physics represents the best of our knowledge, it is also a picture of our world that is fundamentally impersonal, where chance and necessity describe all and leave no room for supernatural agents. It provides no way to anchor spiritual realities in the world as described by science.

This means that in arguments concerning science and religion, the burden of proof has shifted to those who claim that something transcending nature is required to complete our understanding of the world. In today's debate between skeptics and believers, naturalists argue not that current science has all the answers, but that its fundamental picture of the world is the most plausible, and that a naturalistic approach gives us the best prospect of learning more. The religious, in contrast, seek to limit physics, to find some rupture that can signify a divine reality. Some completely reject the scientific approach. More often, however, they try to find gaps in the modern scientific picture of the world.

One way to go looking for gaps is to take an uncompromising stand. For example, some biblical literalists who believe in a few-thousand-year-old universe created in six days propose bizarre physical cosmologies in order to account for this young universe (Humphreys 1994). One of their important pieces of evidence, supposedly a gap unexplained by standard physics, was a so-called quantization of redshifts that suggested the earth was the center of the universe after all. This creationist physics is simply wrong, and recent data indicate that redshift quantization was not real in the first place.

The danger of a "god of the gaps" is that it is too vulnerable to progress in science. Once physicists figure out how to bridge a gap, the God supported by the gap is cast into doubt. Therefore, many among more liberal religious thinkers try and completely detach the spiritual realm from physics: divinity becomes something giving the world meaning and coherence without affecting the physics. This approach, however, risks turning God into a cosmic Santa Claus. If a supernatural force was ultimately in charge of the physical universe, it would be very strange if the science concerned with the fundamental structure of the universe was not able to find any sign of such a force.

So the most interesting of today's arguments for God take a middle path between a god of the gaps and a cosmic Santa Claus. Believers argue that some central ideas in modern physics, not minor details or small gaps, suggest that a spiritual reality underpins physics. They argue that quantum physics brings spirit back into the picture, that a divine design is responsible for the laws of physics, or that physical cosmology has found that the universe had a beginning and hence, a creator. Skeptics respond, arguing that nothing beyond normal physics is called for.

GOD IN THE CRACKS OF CAUSALITY

It is difficult to change the overarching conceptual schemes of physics. But not impossible. In the early twentieth century, first relativity and then quantum mechanics meant physicists had to move beyond classical physics. The Newtonian structure whose clockwork universe was so important to nineteenth-century materialism was found wanting. The new physics, particularly quantum mechanics, inspired religious as well as scientific thought. Conventional theistic thinkers and those closer to occult spirituality alike looked for opportunities to bring spirit back into the world now that Newtonian mechanism no longer reigned supreme.

Newtonian physics remains a very good approximation for everyday objects, and quantum mechanics approaches Newtonian behavior in this limit. Classical physics fails drastically when, for example, explaining how stable atoms can exist. Quantum mechanics does the job, and in doing so, it describes the fundamental nature of physical particles as being radically different than the miniature billiard balls of classical physics.

Physics concerns the states of systems—all the information to be had about a collection of objects—and dynamics, typically equations specifying how a state will change over time. A classical state could be the positions and momenta of a set of particles, and Newton's laws would determine the trajectories of these particles for all times. In practice, we never have exact information about all the particles, and because of dynamical chaos—which means any uncertainties rapidly grow—our ability to predict what a classical system will do may also be severely limited. For classical systems, however, we can always obtain better information about the state of the system and so we can always improve our short-run predictions. Quantum mechanics also deals in states and equations determining exactly how states evolve. The difference is that for a quantum system, we cannot get better information and improve our predictions. There is no such information to be had, and so quantum mechanics can only give the probabilities of observing each of a set of possible outcomes. The result of a quantum mechanical experiment is inherently random.

A common representation of a quantum state is a "wave function," which is closely related to the probability of finding the particle in different locations. Experiments always observe particles, as with photons (particles of light) individually hitting a light-sensitive surface to produce an image. The distribution pattern of photons, however, is wavelike. No one can predict where each individual photon will hit, but with many photons, we get a precise statistical distribution we can calculate by solving a wave equation.

Randomness, then, is fundamental in quantum mechanics. This does not fit comfortably with the rationalist tradition of religious doubt. Nonbelievers, no less than the devout, typically share the commonsense conviction that every event has a cause. In fact, nonbelievers can be even more strongly attached to causality: suggesting that events need not have causes can sound too much like saying they are the result of the inscrutable whims of the gods. Randomness threatens the lawful predictability of nature, along with the rationalist conviction that human reason is able to understand the world. The clockwork universe of classical physics was attractive, as it suggested that *natural causation* was sufficient to account for all that took place. Modern physics, though, is full of random events for which no natural cause can be identified. If so, it can be tempting to think that "random" is just a word we use when we do not know the true causes behind an event. And if the true causes are not to be found within nature, then perhaps they are beyond the natural order.

The breakdown of the classical version of natural causation also inspires more direct theistic arguments. The physicist-turned-priest John Polkinghorne, for example, sees the indeterminacy in modern physics as creating space for divine action in the world (1998). In the context of classical physics, a miracle is a gross violation of physical law, which raises questions about why God would so untidily upset an order that God had determined in the first place. And if God never intervenes, we are back to the religiously unsatisfying option of a remote deistic God. Polkinghorne and others speculate that with indeterminacy, and dynamical chaos amplifying even extremely small fluctuations, the hand of God can again manipulate history. Divine action can work through the cracks in causation, using minute nudges to the dynamics of the world that would be amplified to produce significant results. Creativity can arise from "active information" injected into the world this way. Such divine influence is different than a heavy-handed violation of natural laws; indeed, the divine hand remains physically almost undetectable but still fits in a religious understanding of history.

There is a naturalistic alternative, however, which is a result of taking the randomness in modern physics more seriously. To do that, we need to look more closely at what randomness is.

Consider a sequence of tosses of an ideal fair coin,[1] which goes on indefinitely:

THHTHTTTHTTHHHTTTTHH . . .

Now, contrast this to a sequence that alternates heads and tails:

HTHTHTHTHTHTHTHTHTHT . . .

The second sequence is not random, and it is not random because we can identify a pattern—the alternation—in the series. Mathematically, we can define a sequence produced by flipping an ideal coin as one that is completely patternless. More precisely, no mechanical calculation (or computer program) can produce any infinite subset of a truly random sequence (Chaitin 1987).

If we can find a pattern in a data set, we can provide a short description of the set, such as "alternating heads and tails." Both our ordinary and our scientific ways of reasoning rely heavily on pattern recognition; once we pick out an overall pattern, we can link it to a network of causes in an overall causal explanation. If the set is random, however, there is no short cut. If we want to reproduce the set, we have to list the data exactly as given, as brute facts with no further explanation. In particular, we cannot identify patterns and find a place for the data in a causal account. We cannot legitimately infer any cause responsible for the outcome of individual random events.[2]

Randomness is where explanation comes to an end, and therefore physicists do not declare randomness lightly. In particular, physicists do not give up and say "it's random" just on the basis of ignorance. If we were ignorant of what a coin toss would produce, we could expect repeated tosses to result in anything—a pattern as easily as a patternless sequence. But if we know the toss is random, our inability to tell anything about each *individual* result implies that we know precisely what the *overall* statistical properties of the sequence should look like in the long run. Finding randomness requires knowledge, not total ignorance. When faced with data that looks haphazard, physicists attempt to find patterns, including causal connections, and judge something random only after repeatedly getting nowhere and seeing no prospect of discovering a pattern on closer inspection. Naturally, such a judgment can never be 100% certain; as with any scientific claim, it may turn out to be false. Nevertheless, quantum events appear to be random in just the way an ideal coin toss is, and we can say this with considerable confidence.

The argument that quantum randomness signifies ignorance of nonnatural causes fails by trying to force physics to conform to our everyday intuitions about causes. It overlooks how fundamental randomness is in today's physics, and that we have good reason to expect randomness in our most basic theories—that is where, for now, the explanation ends (Edis 2002).

Indeed, modern physics is full of randomness. For example, the quantum vacuum is often pictured as a sea of short-lived particle-antiparticle pairs randomly popping in and out of existence—not sheer emptiness. This has measurable experimental consequences. Even our everyday lives are touched by randomness, in the form of the disorderly variety of energy transfer known as heat.

So our fundamental theories describing microscopic physics are thoroughly alien to our everyday intuitions. Down deep, our world is a place where particles get created and destroyed with no cause, where interactions are random, and where there is no difference between forward and backward directions of time to distinguish between cause and effect (Stenger 2000). Our familiar, everyday sense of causality is not something fundamental to the universe. Causal regularities of the "the bowling ball struck the pins, causing them to fall" variety emerge in our macroscopic world, and are due to the statistical regularities inherent in large numbers of random events. Insisting that this commonsense notion of causality is fundamental to our world is getting things exactly backwards.

The claim that randomness is a crack in the natural order that makes room for supernatural action is also wrongheaded. Far from allowing miracles without violating physics, such action would require a significant rupture in the present structure of physics. For it is nothing less than a claim that quantum mechanical events are *not* really random. Establishing divine intervention through the cracks would require identifying a pattern in our data that can be linked to the intentions of a supernatural agent. There is no evidence for any such pattern in physics. Perhaps the pattern could be better seen at a higher level, being manifested in miracles or some overall cosmic design. But in that case, speculations about quantum mechanics and causality are entirely beside the point.

QUANTUM MYSTICISM

The occult tradition is not as concerned with a power wholly outside of nature. It tends to conceive of divinity as permeating everything, so that nature is magical at root. Lately, the mystically inclined have sought support in quantum mechanics for their view that consciousness is the fundamental reality (see, for example, Goswami 2001).

One reason for quantum mechanics becoming entangled with mysticism comes from physics itself. In the early twentieth century, as physicists were groping toward an understanding of the weird new physics they had discovered, they naturally tried to make sense of what they found in terms of analogies to classical physics, common-sense convictions about how a reasonable universe would operate, and the philosophical traditions that formed part of their intellectual background. Some of these efforts, even by very eminent physicists, were mystical in tone (Wilber 1984). As quantum mechanics became established as part of the regular equipment of physicists, many of these early speculations were left aside. They stayed alive, however, in circles sympathetic to more occult views of the world.

For example, Heisenberg's famous uncertainty principle has a reputation—outside of physical science—of being an oracular pronouncement about the impossibility of objectivity and the knower determining the state of the known and so forth. To a physicist, it only means that certain properties of a particle such as position and momentum are not simultaneously observable. There is nothing mysterious here, provided we do not attempt to think of microscopic particles like miniature billiard balls with states defined by exact positions and momenta. The quantum state is described by a wave function, and a wave cannot simultaneously occupy an exact position and have an exact momentum. That is all.

There is, however, a genuine problem that has led some to wonder if consciousness has a role in quantum physics. When we make a measurement, we get a single value. For simplicity, say we can get either result A or result B when we perform a certain experiment. The problem is, quantum states are generally like $a\psi_A + b\psi_B$, where $\psi_{A,B}$

refers to the states corresponding to result *A* or *B*, and *a* and *b* are numbers that determine the probability of getting each result. After measurement, we come up with either *A* or *B*, not a combination. This means we lose information, throwing away any knowledge about *a* or *b*. But the equations of quantum mechanics preserve information, and measurement, like everything, should be a quantum process. Why does it appear to be an exception? Some have suspected that a definite *A* or *B* is determined only when an observer enters the scene, that consciousness makes the original state "collapse" to give us either ψ_A or ψ_B after a measurement.

This is still not an issue that has been resolved to every physicist's satisfaction. Even so, some major pieces of the puzzle are in place. It appears likely that we get a classical macroscopic world with definite *A*'s and *B*'s for reasons similar to the way we get causality out of a microscopic physics full of randomness.

Imagine a movie of two billiard balls colliding, bouncing off one another. Now, if we ran the movie backward, we would again see two balls approaching one another, colliding, and bouncing off. In fact, we cannot tell if a single-collision movie is being shown forward or backward. The equations describing the motion, whether they are Newton's laws or quantum mechanics, remain the same under the mathematical operation reversing the direction of time.[3] But now take a movie of two cars colliding and run that backward. We can immediately tell it is backward—haphazardly crumpled up piles of metal and plastic do not spontaneously assemble themselves into two well-organized functional machines. Yet, the interactions of the enormous number of particles making up the cars is described by microscopic equations as time-reversible as those for the two billiard balls. What is going on?

Numbers make a difference. If we just looked at a movie of a break shot starting a pool game and ran it in reverse, we could also see a difference (see Figure 2.1). It is possible for 16 billiard balls to all come in from different parts of the table and have exactly the right states of motion for 15 to assemble themselves into a triangle and kick out the sixteenth toward the other end of the table. But the probability of such spontaneous order is *extremely* low. Even the slightest variation in the initial states of the balls will prevent the triangular order and result in a shapeless mess.

With the about 10^{20} to 10^{25} particles in everyday macroscopic objects, obtaining spontaneous order by haphazardly throwing particles

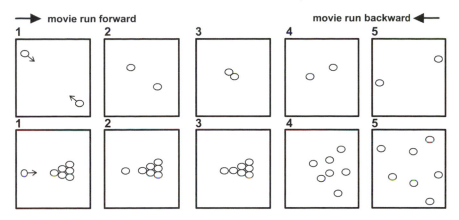

Figure 2.1 The upper row shows frames from a movie of two billiard balls colliding. The movie run backward would not look odd in any way. The lower movie, of the break shot, would show an extremely improbable event if run in reverse. Yet, the physical laws underlying both movies are the same, and do not distinguish between forward and backward directions of time in either case. (Ricochet Productions)

together becomes virtually impossible: the number of disordered states completely overwhelms the few ordered states. It also becomes impossible to calculate how objects behave by solving the equations for every particle. So physicists look for macroscopic variables—like pressure or temperature for a gas—that describe aggregate statistical properties. The equations for how these variables will behave can be derived from the microscopic equations, but these macroscopic equations turn out not to have the time-reversibility property of microscopic physics. Heat, for example, will spontaneously flow from hot to cold, but not the other way around. Going to a macroscopic description means losing information over time, and so time has a forward direction in the macroscopic world (Edis 2002, chapter 2; Price 1996).

Quantum measurement is similar. Measurement is a macroscopic, complicated process. Due to disordered interactions with their environment, quantum systems undergo "decoherence" and evolve toward macroscopic states with classical properties (Zurek 1991). The conscious observer is involved only to the extent that she is a macroscopic object interacting with the quantum system.

Quantum physics is weird. It is impossible to visualize in commonsense terms. Nothing in the quantum world, however, calls for mysticism or any special role for consciousness. Physicists continue to dispute the role macroscopic decoherence plays in obtaining classical measurements, but even those inclined to minimize its role do not treat observation as a magic act.

COSMIC HARMONIES

Within randomness is a bad place to seek supernatural agents. Another tradition religious arguments have often drawn on, however, is the notion that our universe seems particularly well ordered. The laws of nature themselves suggest to many thinkers that a cosmic intelligence has set the course of our world.

Such ideas have ancient philosophical roots. Greek rationalism never mixed well with the popular gods, but philosophers often reconciled reason with a suitably reinterpreted religion by infusing reason itself with spiritual qualities. Our living in a lawful world instead of a formless chaos suggested a divine intelligence had made it so. Furthermore, contemplating the order of the heavens could be the best way to approach the transcendent realms. Philosophical nonbelievers themselves were often influenced by this tradition; they just emphasized the sufficiency of reason and saw no need for anything beyond.

Today's physics can still inspire such thoughts. The popular image of theoretical physics is that of geniuses with bad hair investigating the sublime symmetries of nature, thinking their way toward elegant laws of physics where beauty becomes truth. Many physicists strive for a "theory of everything," a set of equations that would fit on a T-shirt and describe the fundamental structure of the universe. And the notion of symmetry is central to how physicists construct fundamental theories today. Just as momentum is conserved in a universe that is symmetric in space—location makes no difference to the laws of physics—physicists try to generate all principles of physics from the symmetries of exotic mathematical spaces. Theoreticians seeking fundamental symmetries, armed with notions of mathematical elegance as a guide, do not appear too different from philosophers contemplating eternal harmonies of the heavens and thereby touching more transcendent spheres of existence.

Even if all this were correct, its significance for religion would be murky. Religiously inclined thinkers see hints of the transcendent in the symmetries of nature. Those with a more skeptical bent, however, note that fundamental physics brings forth an elegant but impersonal

order. Bringing supernatural agents into the picture, with all their unavoidably personal qualities, would destroy the spare elegance that
makes physics so attractive. In any case, this dispute misses the point.
It loses sight of what the physical significance of a set of equations on
a T-shirt would be.

Imagine that a theorist came up with the holy grail of physics: an
elegant equation that showed how to obtain the long-sought, full theory of quantum gravity; unified all the basic forces; and so captured
all interactions between fundamental particles in a single account. She
would certainly win a Nobel prize, but curiously, she would also do
little toward explaining the details of our world. Perfect symmetry is
featureless in the same way a sphere looks the same from all angles.
An equation on a T-shirt has very little information content; it actually
would say very little about our particular universe. No one could start
from that equation and calculate what our world actually looks like.

How, then, can we get a messy, information-rich, detailed universe
out of so blandly symmetric laws of physics? Much like the way we
get magnets, it turns out.

Our familiar magnets are essentially large collections of atomic-size
magnets that interact with each other in such a way that the microscopic magnets align with each other. At high temperatures, the haphazard thermal motion of the micromagnets overwhelm their mutual
interaction. So they point in random directions and cancel out each
other's effects, giving an overall average magnetization of zero. This
is, in fact, what we would expect from the equations of electromag-
netism. These equations are rotationally symmetric, not favoring any
direction in space over any other. The high-temperature magnet, having zero magnetization, also does not pick out any direction in space.
As we cool the magnet down, however, the micromagnets become less
agitated. There comes a temperature below which the mutual interactions of the micromagnets dominates. Then the micromagnets
arrange themselves to point in the same direction, giving a nonzero
average magnetization for the overall magnet. This state is less symmetric than the equations describing magnetism, because it singles out
a direction. The direction of the overall magnetization is completely
arbitrary: it is selected at random (see Figure 2.2).

This process of "symmetry breaking" is very general, and it also describes what happened in the early universe. Immediately after the
big bang, the universe was extremely small, and its temperature was
enormous. At this point, everything was symmetric and unified—and

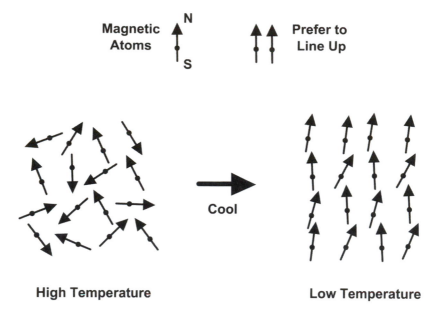

Figure 2.2 An ordinary magnet is made up of many micromagnets that interact with each other such that they prefer to line up in the same direction. But at high temperatures, this weak interaction is overwhelmed, and the magnet has no overall magnetization. When it cools down, the symmetry is broken, with the micromagnets tending to line up in a common direction. This direction is arbitrary. (Ricochet Productions)

not very interesting. This symmetry indicates a fundamentally simple universe, just like it is simplest for the laws of magnetism not to have a preferred direction in space. A preferred direction would call for an explanation, while symmetry does not (Stenger 2006). Complexity arrived soon, however. As the universe expanded and cooled, its symmetries broke, starting with the various forces becoming distinguished from one another. In other words, our particular universe emerged out of a cascading series of accidents. The physics we know comes not just from the symmetries on a T-shirt, but from those random events of symmetry-breaking that led to a nonsymmetric, complicated low-temperature physics. We can easily imagine universes that ended up looking very different from ours; indeed, many theories in physical cosmology today indicate a multitude of separate universes with widely varying physics.

So we should not exaggerate the significance of fundamental symmetries. These, the most basic laws of physics, do not alone tell us what the world looks like. They only say what sort of dice to roll to generate the world. Our universe operates by a combination of elegant laws and sheer randomness—chance and necessity. In fact, symmetry and randomness go hand in hand. An ideal fair coin is symmetric with respect to heads and tails, not favoring one over the other. And it is just this symmetry that gives us the equal probabilities of heads or tails in an ideal coin flip. Laws and randomness in modern physics are as inseparable as two sides of the same coin (Edis 2002).

FINE-TUNING THE UNIVERSE

If modern physics is ambitious, nowhere is this more obvious than in the fact that today we can do cosmology scientifically. Physical cosmologists deal with all of space-time as a single geometric entity, they speak of the creation and death of universes, and they try to explain what astronomers observe billions of light-years away. And so cosmologists trespass on prime religious real estate.

Physical cosmology can inspire a purely secular awe, but it provides an opportunity for religious arguments as well. If looking for transcendence in the symmetries of fundamental physics is misleading, we can still ask why we are so lucky to have had just the right sort of accidents that led to a universe which can support life and intelligence: us humans!

There is a long tradition of arguing that the physical coincidences that allow us to exist are just too improbable, and so an outside intelligence must have intervened to fine-tune the universe for our sake. A couple of centuries ago a similar argument featured the fact that the solar system is stable, going on to observe that stability called for a very specific arrangement and motion of the planets, and that this was just too improbable to have happened by accident. A stable solar system being necessary for human existence, our welfare must have been the reason for it all (Greene 1959, 83). Progress in physics, such as a good theory of planet formation, tends to dispose of such arguments. There is also a religious tendency to try and have it both ways. If we appear unlikely, religious thinkers say an intelligence must have intervened to produce us. On the other hand, if it looks like nature is such that our existence is no surprise, then we should admire the intelligence who decreed that the laws of nature make sure we exist. This "heads belief wins, tails nonbelief loses" approach suggests that the fine-tuning argument lacks in substance.

Nevertheless, fine-tuning arguments in favor of a God raise some real physical questions, even though they try to prematurely close off investigation in favor of a "God did it" explanation. Our current physics contains many constants that are not fixed by any deeper theory. So we can easily imagine universes where these constants take on different values. It turns out that if we vary certain parameters of a universe such as its initial expansion rate, relative strengths of fundamental forces, the mass ratios of elementary particles and so forth, we get universes that quickly implode, or have all matter bound up in uninteresting states, or in some way or another prevent complex life and intelligence from ever arriving on the scene. Indeed, even very slight variations of the parameters from their present values give us dead universes. So, if the roll of the dice that led to our universe produced such an improbable result—and a clearly special sort of result—we may well suspect the dice were loaded.

Physicists naturally want to address the fine-tuning problem. And the first step is to see how real the fine-tuning is. It appears many claims of fine-tuning and "anthropic coincidences" are exaggerated (Klee 2002). We should not ask for the likelihood of obtaining life as we know it; that is too narrow a set of outcomes. It is more meaningful to ask for a universe that is large, long-lived, and can support a diversity of environments, including some where complex chemistry and biological evolution can take hold. Varying the fundamental constants and examining what sort of universe would result shows that a universe that can support complexity is not as grossly unlikely as claimed by some fine-tuning enthusiasts (Stenger 2004). The dice may not be so seriously loaded after all.

There is also the question of what any fine-tuning might signify, even if it were real. After all, fine-tuning problems are not new to physics. In cosmology alone, many different fine-tuning problems have been encountered and resolved over the years. They are generally a sign that some new physics is around the corner. For example, inflation, one of the most fruitful ideas in recent physical cosmology, emerged as a solution to some vexing fine-tuning issues that beset earlier models of the big bang and the expansion of the universe (Guth 1997). In all likelihood, current fine-tuning problems will be resolved the same way, by bringing in new physical ideas that produce better cosmological models. After all, we already know that current physics is not adequate to completely address cosmological questions. This is because gravity is the most important force at cosmic distances, determining the very shape of space-time. And we do not yet have a complete theory of quantum gravity. So claiming fine-tuning is like claiming loaded dice when we do not have an adequate description of what sort of dice were thrown in the first place.

Another idea that deflates the force of fine-tuning arguments is the possibility that ours is not the only universe. If ours is one in a large population of universes, then some unusual universes are likely. An improbable result does not lead us to suspect a loaded die if a multitude of dice were cast.

At first, multiple universes seems like a peculiar idea. It could even seem to be an arbitrary invention to avoid genuinely solving a puzzle, if it were not for the fact that multiple universes arise naturally in many theories that have independent support. It is not unusual today for cosmologists to speak of universes as bubbles in a "false vacuum," of many "baby universes" inflated into existence, or even to speculate about how the creation of a black hole in one universe could start a new universe connected to the original. There is even Lee Smolin's (1997) intriguing idea of obtaining something like Darwinian evolution in a population of universes reproducing through black holes. In such cases, it would be a disruption of the structure of physical theory to deny the existence of other universes (Davies 2004).

At present, none of these approaches to fine-tuning can be said to be *the* solution. Cosmology is a young science. It has a few very solid results like the fact that our universe has been expanding from a denser, hotter past state, but otherwise it is a wide-open field where the theoretical ideas in play are unavoidably somewhat speculative. Even so, fine-tuning looks like an ordinary physical problem, solvable in physical terms. Suggesting "God did it" as an explanation is just asking for this latest god of the gaps to be superceded one day as physics continues to advance.

In fact, "God did it" turns out to be just about useless if considered as a serious explanation for fine-tuning. After all, we can ask not just what happens if we vary a few physical constants, but look at the full range of possible worlds with different physical laws or even entirely different ways of operating. Then we can ask why we do not live in a world where the reality of a spiritual realm was a touch more obvious. Many scriptures and religious works of literature imagine just such worlds; some even assert, mistakenly, that we inhabit such a world. Why not Dante's universe, or the cosmos of the Quran? Why not a world with an explicit moral order in nature, instead of impersonal symmetries and randomness at the bottom of it all? Why are we fine-tuned to *this*?

Theologians can be ingenious in producing excuses for their gods, but usually where the argument ends up is that the ways of the Lord are mysterious. Maybe so. But if we have no idea, independent of what we already know about our universe, of what kind of universe the gods might have wanted to create when they set nature up and fine-tuned it, then "God did it" can hardly be an explanation to compete with even the most speculative physical idea.

A MOMENT OF CREATION?

Physical cosmology inspires another popular argument for a creator God. Standard big bang theory indicates that the universe had a beginning, as the universe exploded into being from a very hot, very dense point. Even today, we observe that the universe continues to expand, and we can see the cooled-off remnants of the initial explosion in the cosmic microwave background. Before the big bang, there was nothing. This, many have thought, sounds very much like a universe created out of nothing, just as in the theologies of the monotheistic religions.

In fact, nonbelievers have often been uncomfortable with the big bang account of the universe. Here, physical cosmology intrudes on a long-standing dispute between Greek philosophy and revealed religion. Orthodox Jews, Christians, and Muslims insisted that the world was created, since only God could be eternal. The philosophers, though they affirmed transcendent realities and an abstract ultimate divinity, ended up thinking that the world was eternal. The universe depended on God, it was always emanated from God, but nevertheless it had been around forever. Religious doubters agreed with the philosophers that the universe was eternal; after all, an eternal world can more easily be thought not to depend on any divinity. So the claim of an eternal world became a marker of nonbelief, or at least unorthodoxy.

Coming from this tradition, nonbelievers have often been suspicious of any suggestion that the world had a beginning. This suspicion has further been fueled by rationalist beliefs that everything has a cause, and that natural causes alone account for all in the world. In that case, if the universe *began*, it must have a cause, and it looks like that cause stands outside nature (Craig and Smith 1993). Crank writings denouncing the big bang theory typically have one of two sources: fundamentalists who dislike the 14-billion-year-old universe, and atheists who dislike the idea that it has any age at all.

This whole debate, however, is misconceived. Philosophical positions concerning the eternity of the world took shape before modern physics, and relied on commonsense assumptions about the nature of time and causality. The picture provided by modern physics is considerably weirder.

A common picture of the big bang is that 14 billion years ago, all matter originated in an explosion in empty space. This is mistaken. In the theory of general relativity, which is the framework for the standard big bang, space, time, and matter/energy are intimately connected. Gravity arises from the way energy warps space-time. Time and space, then, need not stretch out to infinity. Just like the Earth is not flat, space-time as a whole can be curved.

The standard big bang is where time begins, and within the context of general relativity, asking what happened in the time before the big bang is like asking where is the next town north of the North Pole (see Figure 2.3). Still, according to general relativity the big bang is a "singularity," a boundary of space-time, and it might seem to be a point of creation. Physically, however, what comes out of the big bang is random. We cannot infer any cause beyond a singularity any more than a cause behind a quantum fluctuation. Something else, like a design argument, is necessary to show that what happened was not random before speculating about creators.

In any case, though the standard big bang theory is an excellent description of the early universe, at the big bang itself the theory has to break down. General relativity is not a quantum theory, and at such extreme circumstances quantum gravitational effects should dominate. Unfortunately, we do not have a full theory of quantum gravity. But we can still look at some current cosmological ideas that approximately combine gravity and quantum mechanics (Hawking 2001). In all cases, the big bang is no longer a point of origin.

One example is James Hartle and Stephen Hawking's no-boundary cosmology. This proposal preserves much of the standard big bang picture, but the big bang is no longer a boundary. It cannot be singled out as a special point in space-time. Without the singularity, the "be-

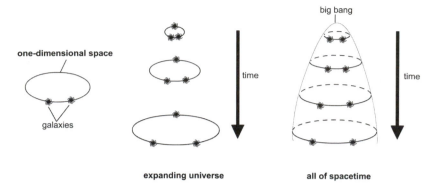

Figure 2.3 To visualize the big bang, imagine that space was one-dimensional, like a circle, instead of our three-dimensional world. An expanding universe means the circle's size grows as time passes. All of this "1 + 1 dimensional" space-time is then a curved surface, and the big bang is the point where the size of the universe becomes zero. Asking what comes before is like asking what is north of the North Pole. (Ricochet Productions)

ginning" of time will differ for different observers. So we are left with a self-contained universe that is nevertheless finite in time. This scenario is not the most popular option among cosmologists, but it illustrates how an eternal universe and a universe with a beginning are not the only two options. Once we drop our commonsense view of time and understand it within modern physics, other possibilities arise.

Another view of the big bang is inspired by the theory of inflation, which was introduced to solve problems such as the fine-tuning of the initial expansion of the early universe, and the fact that space looks much the same in all directions even though very distant regions could not have had any physical effect on one another since the big bang. Inflationary theories hold that immediately after the big bang, a part of space expanded much faster than the speed of light for a while. Trying to make this idea work naturally led to the notion of inflating other universes into existence. Andrei Linde's scenario of an eternal self-reproducing universe has many connected universe-bubbles inflating, with different low-temperature physics in each baby universe. Here too, the universe is neither straightforwardly eternal nor with any beginning, as each universe-bubble has its own separate time dimension (Stenger 2000).

More recently, physicists have been focusing on string theory and on loop quantum gravity as paths that promise to lead to a full account of quantum gravity. Naturally, such ideas have implications for the big bang—both approaches produce cosmological models with an infinitely long history before the big bang. In string theory, "strings" rather than point particles are the elementary objects (Greene 1999), and strings resist collapsing to a point beyond a smallest length scale. So string theory allows cosmologists to speak of an eternal universe, in that a time dimension can be extended infinitely backward beyond the big bang. For example, in Maurizio Gasperini and Gabriele Veneziano's scenario (2003), the big bang marks a transition between a pre-big bang state where matter inside a black hole collapsed until reaching the string-theoretical limit, on which it rebounded in an outward explosion. Interestingly, this scenario also calls for multiple universes generated by similar processes.

Cosmological scenarios can be hard to test, but not impossible—cosmologists routinely rule out models when their predictions do not match observations. And with better space-based telescopes and entirely new instruments such as gravity-wave detectors about to come into use, models concerning the early universe will increasingly be put to the test. Furthermore, we can expect surprises. Recently, observational data indicated that the universe is expanding at an accelerating rate, which no one expected. This discovery has raised many questions, opened up new directions of research, and any explanation for it will have wider cosmological implications.

Even though cosmology is rapidly changing, and entirely new scenarios are likely to appear in the coming decades, its significance for religious nonbelief is clear enough. As is always the case in modern physics, cosmology is making good progress in a naturalistic framework, without any gods. That much fits what nonbelievers would expect. However, skeptics no less than believers have been slow to recognize that the ancient debate about whether the universe is eternal or whether it has a beginning is no longer relevant. Modern cosmology has changed what we mean by time, space, and universes.

PHYSICS AND NATURALISM

Practically no physicist would be rash enough to say that the present state of play in physics absolutely rules out any supernatural agent. If any would make such a claim, it would be easy to enough to point out that physicists, like any other brand of scientists, have no "theometer" (Scott 2001). No one can construct a god-detector, walk into the lab, and report that they have found none. And there is no shortage of religious physicists or of theologians interested in physics, and the strange ideas of modern physics inspire many a speculation on the nature of the God they assume is behind it all.

Nevertheless, it is also true that in physics today, God is a wholly unnecessary hypothesis. And physics is an ambitious science that has a way of charging in trying to fit everything into a physical account of the world. It is significant that a science that is so fundamental, so successful, and so wide in scope, should end up with a purely naturalistic picture. Physicists have no theometer, but then again, they have no big bangometer either. Theoretical entities that are supposed to anchor a broad theory and help explain a wide range of phenomena are rarely detectable in any simple fashion. There is no single observation, not even the cosmic microwave background, which is a direct, unambiguous signature of the big bang. All lines of evidence for the big bang can be explained otherwise. What makes it virtually certain that our universe was in a much denser, hotter state about 14 billion years ago is that this scenario ties together many independent lines of evidence, and that it takes a central role in productive cosmological theories. Similarly, a physics-transcending spiritual entity could, conceivably, have had a prominent role in a successful overarching theory making sense of our universe. Attempts to find signs of a God in quantum phenomena or in fine-tuning and so forth grope toward just such an overarching account where a divine reality ties everything together. None of these attempts work. And therefore God the creator runs the risk of becoming dispensable, a cosmic Santa Claus.

So nonbelievers today often claim support from modern physics. Physics is not greatly relevant to the moral or purely philosophical critiques of religion that continue to be major motivations for nonbelief. Nevertheless, the work of physicists remains a leading support for the dominance of naturalism in science today.

This is not to say that physics and the mainstream tradition of nonbelief have no points of friction. Besides being too closely wedded to a naive version of an eternal universe, nonbelievers have also often been slow to come to terms with the fundamental randomness of our physics. This is curious, as genuine randomness undermines any way of understanding the universe in purposive, spiritual terms. After all, blind chance is just about the only thing more mindless than impersonal natural causality. But chance brings the unwelcome prospect of the world being a formless chaos where anything goes, and rationalists worry that we cannot even count on the sort of regularity that would allow us to do science.

Randomness, however, is very different than sheer ignorance. And "formless chaos" is a commonsense intuition that has to be set aside to understand how the randomness in modern physics works. Not anything goes, and there is no threat to science. Religious doubt has a long intellectual history, and many a nonbeliever has brought up chance as an alternative to divine purpose. Broadly speaking, any naturalistic view must hold up chance and necessity against divine providence. Modern physics shows just how our world is a place of chance and necessity, where no providence is visible. It enriches our understanding of how chance and necessity are inseparable, two sides of the same coin. And in doing so, physics continues to seriously challenge religious ways of viewing the world.

NOTES

1. It is important that this is an ideal coin, not an actual coin. Real coins are macroscopic, classical objects. We can always in principle obtain more information about the state of the system and its environment and so trace the complicated series of causes that lead to a heads or tails result. Quantum events are like the mathematical ideal, where each toss is entirely independent and where there is no further information to be obtained that can improve our ability to predict the result of any individual toss (Edis 2003a).

2. Some physicists adopt interpretations of quantum mechanics where random events have *unknowable* causes, such as Bohmian hidden variables. But, these causes do no real work—they do not get rid of the randomness but only displace it from the dynamics to the initial conditions. Thus, the implications of quantum randomness remain the same with such interpretations as well.

3. To be completely accurate, physics is invariant if we not just reverse time but also replace everything with its mirror image and replace all particles with their antiparticles. But this detail does not affect the discussion. Also, friction has to be ignored in the billiard ball example, since friction is a macroscopic, multiparticle interaction.

Chapter 3

Darwinian Creativity

ACHIEVING COMPLEXITY

Physics today has no use for supernatural forces, either for under-standing the smallest constituents of matter, or for investigating our cosmos over the grandest sweeps of space and time. Still, even after physics came into its own, many religious thinkers continued to see spiritual forces shaping the world. Newtonian physics explained a lot, but it was difficult to see how it could account for complexity. How does a delicate order arise out of just physical objects bouncing around? In particular, the intricate functional complexity of life was a stumbling block for anyone dreaming of explaining the world without any gods. Life just *looks* designed.

Outside the scientific mainstream, such challenges are still heard today. There are, after all, some good questions to ask. Life appears to embody a special kind of order that is not like the statistical regular-ities shown by a collection of randomly behaving particles. How does this happen? Intuitively, we feel that there is something special about life—our brains are wired to treat living things, which move under their own volition, different from inanimate objects. To these living beings, we apply folk biology, classifying them under different species. We usually think that members of a species share a particu-lar essence, and that therefore one species cannot transform into an-other. Furthermore, the adaptations that life forms show to their way of life are so detailed, so wonderful as to excite awe. We are best acquainted with functional adaptations in human artifacts, which are typically designed to fit a particular purpose. And so often the commonsense view of life is a creationist one: life does not just evolve on its own; it has to be created, intelligently designed.

Against the view that complex order trickles down from above, imposed by intelligence, scientific naturalists propose that order bubbles up from below. Before modern science, nonbelievers frequently relied on the ancient Greek notion that everything was composed of indivisible atoms and speculated on how natural processes could assemble atoms into complex bodies. This approach, however, was not entirely convincing. Atomists such as Lucretius (1995) promised a unified view of the world: all were made of atoms. But when it came to details, they did not successfully explain much. To account for how some material things could live and feel, atomists could only propose that the soul itself was composed of particularly light and fine atoms and that somehow their interactions could grant humans psychic qualities. Naturalistic approaches have to try and account for life and mind in terms of the interactions of mindless, lifeless physical particles. Atomists could only gesture toward how this was supposed to happen.

So well into the nineteenth century, the most ambitious vision of physics, the idea that we could unify all our knowledge of the natural world with our understanding of physical science, had a huge gap with which to contend. Complexity, especially life, looked like it was beyond mere physics.

Today's picture is considerably different. Although the ancient idea of a mysterious life force that animates living beings remains popular, continually resurfacing in science fiction and in New Age spiritualities, it has long been discarded from biology. Many conservative religious believers remain convinced that basic life forms are immutable, and that complexity is a sure sign of design, but in biology such views do not command any more respect than belief in a flat Earth.

Modern biologists continue to do natural history, describing the amazing variety of life. Like their colleagues a few centuries ago, they also study animal behavior, ecology, and anatomy. Today, however, they also probe the intricacies of microbiology, genetics, and biochemistry at a depth not comparable to anything imagined in the nineteenth century. Theoretical biologists examine evolution as a physical process, connect genetics to concepts of information and complexity derived from physics and computer science, and even work on "artificial life" sustained in computer environments (Ward 2000). Life scientists ask how life might arise on other planets, engineer DNA to create programmable living machines, and debate evolutionary explanations of self-sacrificing behaviors. Biology has become a mature science, comparable to physics in its theoretical and experimental sophistication. Like physicists, today's biologists build complex mathematical models and hone them with computer simulations as well as test them in the real world. So while advancing, biology has forged closer connections with physics and chemistry. These historically distinct scientific disciplines now shade continuously into one another. Biology has also developed conceptual schemes, such as Darwinian evolution, which describe the world in terms often offensive to common sense.

The intellectual centerpiece of modern biology is evolution. Living things on our planet are related to one another by common descent: a cat and a cockroach share an ancestor some time in the remote past. Without the framework of historical connections provided by evolution, much of biology would be little more than the sort of stamp collecting typical of old-fashioned natural history. Just recognizing the bare fact of common descent, however, would not have been enough to make biology a mature science on a par with physics. Having a theory—a detailed explanatory framework complete with testable mechanisms—is the important thing. Modern biology recognizes many factors that shape the history of life, including essentially accidental processes such as genetic drift and mass extinctions. But the most interesting questions have always had to do with adaptation, functional complexity, and the sheer creativity seen in the abundant variety of earthly life. And in modern biology's answers to such questions, the heavy lifting is done by Darwin's mechanism of variation and selection. Indeed, the Darwinian mechanism appears to be critical for achieving functional complexity in a world made of otherwise lifeless, mindless particles. The crucial piece to the physicists' puzzle of how to achieve complexity was supplied by the biologists.

Darwinian evolution is comparable with the best theories in physics in its abstract elegance. It starts with a population of replicators—objects able to copy themselves and thereby propagate information into the next generation of replicators (Dawkins 1986). The population should not be uniform: there must be some variability among the replicators. Furthermore, while the replication process must take place with reasonably high fidelity, it must not be perfect: *new* variations must be introduced into the population with each generation. These variations must be *blind*, in the sense that they should not be designed to work better or be biased in any forward-looking way. The best way to get blind variation is to have some completely random element in the variations. Finally, replication must proceed at different rates, with success or failure at least partly depending on the information being replicated. In these conditions, the stage is set for Darwinian evolution. Not all variants will be equally successful in replicating, and those that replicate better will be better represented in successive generations. Random variation introduces novel features into the population. If dumb luck produces a new variant replicator that does slightly better, it will have a good chance of flourishing. Though the Darwinian mechanism is blind, and does not favor the complex over the simple, *if* a complex adaptation does well in a particular local environment, the Darwinian scenario can describe how it can arise and how it will enjoy a selective advantage.

Darwinian explanations are most common in biology, where they help account for adaptation, novel structures, and the branching structure of descent seen in the history of life. However, it can be applied to replicators other than those that propagate genetic information. Abstract replicator theory and Darwinian ideas come into play today in chemistry, physics, artificial intelligence, and brain science—anywhere a population of replicators exists or functional complexity needs to be explained.

Like modern physical theories, Darwinian evolution combines chance and necessity. And the effect of evolution is to place creativity squarely within the natural world. So unsurprisingly, Darwin has become an icon among naturalistic nonbelievers. Moreover, evolution more directly relates to religious concerns. Biological evolution is greeted enthusiastically by skeptics and treated with suspicion by conservative believers, more so than any of the challenges to spiritual views arising from modern physics.

THE PHYSICS OF EVOLUTION

Biologists have abundantly confirmed that all life forms are related through common descent. The fossil record, the geographical distributions of different species, the nested hierarchical structure we get when classifying species, peculiarities like the "panda's thumb" (Gould 1980) that arise from the historical constraints on evolution, the evidence of molecular biology, all converge to make evolution a clear fact as far as science is concerned. The mechanisms that drive evolution are more open to debate, but there the scientific discussion concerns questions like the relative importance of various mechanisms in shaping the precise history of life, the pace of speciation events, or whether evolution should best be approached from a gene-centered point of view. Purposeful, forward-looking influences on evolution are not considered serious options.

Still, how the consensus of biologists relates to the world described by physicists needs to be fleshed out. Naturalists strive for completeness: evolution should, they expect, fit smoothly with what is known in physical science. In particular, evolution should not require even indirect action by supernatural forces.

How evolution fits in with physics is not, however, obvious at first glance. The most striking result of biological evolution is that starting from what must have been very simple chemical replicators, there came the amazingly complex structures of bacteria, eukaryotic cells, multicellular creatures, and mammals such as ourselves. That a blind physical process should produce complex order is strange enough, but it is especially strange considering that according to the famous second law of thermodynamics, the entropy of the universe, which is a measure of its disorder, can only increase over time.

Microscopic physical laws do not distinguish between forward and backward directions of time, but in our everyday experience, the direction of time is obvious. When dealing with macroscopic systems, we can only keep track of variables describing statistical properties of very large numbers of particles. Working with a macroscopic description means accepting a continual degradation of our information about a system. There are vastly many more ways in which a macroscopic system can be disordered than ordered in any meaningful way. So we expect that disorder should be more likely, and that ordered systems will tend to disorder over time. Entropy grows. So, as generations of creationists have asked, how is evolution possible? How does the information specifying the incredibly complex structures of living things arise if it is not injected into the universe by an outside intelligence?

The answer is complicated. To get there, we have to look at local reductions in disorder, self-organization, thermodynamics in systems driven away from equilibrium, and how the Darwinian mechanism works in such circumstances.

First, the second law of thermodynamics allows local reductions of entropy, as long as the total entropy of the universe increases. An animal can grow and heal and otherwise decrease its disorder, but only so long as it can discard a good deal of waste and cause even greater disorder in the rest of the universe. Spontaneous ordering cannot happen in systems that are isolated from their surroundings, but both individual life forms and the biosphere of our planet are open systems. Energy flows through them, in the form of food consumed and work done on the environment, and the sunlight that reaches the surface of the earth and is reradiated into space at different wavelengths.

Second, simply having an open system with energy flowing through is not enough. For example, energy flow in the form of ionizing radiation would disrupt chemical order rather than sustain it. Under the right circumstances, however, a spontaneous ordering known as self-organization takes place in open systems. Consider Bénard cells, which form when a thin layer of water is uniformly heated from below. Heated water expands, becoming less dense and rises toward the surface. Colder water then replaces it below. Water molecules rising and falling independently is not the most efficient and stable way of sustaining this exchange of hot and cold water. Instead, we get hot and cold streams organized into macroscopic flows, where cold water falls in hexagonally arranged columns and where the hot water rises along the edges of these hexagons. So the energy flow through this particular open system spontaneously organizes a macroscopic order, in the form of clearly visible hexagonal cells (Shanks and Karsai 2004).

Self-organization, in much more intricate forms than simple Bénard cells, is important for biological function. For example, genes are instructions for assembling proteins, not blueprints that specify exactly how an organism should be built. Building a body depends on self-organizing processes triggered by genes—the genetics does not completely determine the end product. So genetically identical clones of an animal can be significantly different—for example, have different coat colorations—due to variations in developmental environments that channel self-organizing processes in different directions.

Self-organization can only take place in systems far from thermodynamic equilibrium. A third question, then, is how to get away from equilibrium. After all, thermodynamics says that physical systems tend toward increasing disorder and an overall equal temperature. In time, we should end up with a featureless, dead soup of particles randomly bouncing around. So evolution requires a way to make open systems with energy flows, keeping them far from the deathly uniformity of equilibrium.

There are many ways to get far from equilibrium, but one of the most important is to have a system that does not have a fixed number of possible physical configurations. If the set of possibilities expands in time, both entropy and spontaneous order can increase together. For example, we live in an expanding universe. Around the big bang, the universe had close to maximum entropy, but as everything was compressed into such a small space, the maximum possible entropy was not too large. After the big bang, the expansion of the universe meant the maximum possible entropy not only grew, but it grew faster than the actual entropy of the universe. Hence, our universe was driven away from equilibrium, allowing order to form (Edis 1998b). Biological evolution proceeds similarly (see Figure 3.1). As time goes by, the maximum possible diversity of life forms increases faster than the actually realized diversity (Brooks and Wiley 1988).

All of this sets the stage for evolution, but a fourth step remains. We need a mechanism to create information. Self-organization far from equilibrium can give us impressive and intricate structures, but these are physically constrained. Life requires more flexibility. Consider the genetic code. Just because it is a *code*, it embodies a special sort of order. Codes can express gibberish just as easily as meaningful information, just like a string of letters can be a meaningful sentence or a jumble like "elkfgeufgujgqhwfiggoigwgeh." If strings of letters were physically constrained to give a particular sentence all the time, they would be useless as a code, which should be able to express a vast variety of messages.

This is where variation and selection comes in. Random variation goes beyond any physical constraint, producing truly novel genes. Most variations do not help, degrading the ability of the bearers of the new genes to replicate themselves. But some do help. Selection picks out those DNA sequences that are not gibberish, in the sense of leading to a viable organism that reproduces well. So Darwinian evolution creates information, constructing genuinely new, complex structures even though it is a mindless, directionless process.

A Darwinian process cannot take place in just any environment, just as not every energy flow leads to self-organization. If, for example,

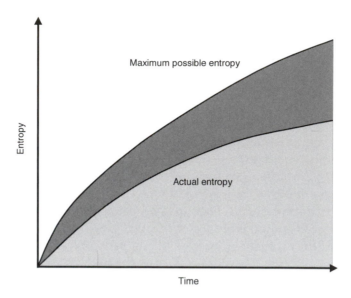

Figure 3.1 A system where the maximum possible total entropy (disorder) grows faster than the actual entropy is driven ever further from equilibrium. The distance between the two entropy curves indicates an increasing opportunity for order to spontaneously form.

small variations in DNA would always produce haphazard, sometimes catastrophically large changes in the ability to reproduce, variation and selection would not be much good for accumulating information about what works best. So we expect that not too many environments would be suitable for originating and sustaining life. Life is not likely to happen in the core of a star or in interstellar space. But in a large, long-lived universe with plenty of diversity of physical and chemical environments, evolution has a good shot at getting started somewhere, sometime. Our universe is such a place.

Evolution, then, is not just a bright idea belonging to biologists, nor just an explanation of the fossil record. Evolution is central to a detailed picture that modern science has put together, of how order arises from chaos.

EVOLUTION AND RELIGION

Complex order, in the modern naturalistic scenario, emerges through chance and necessity, from a world of fundamentally lifeless and mindless physical processes. This is not just a philosophical principle asserted by nonbelievers disgruntled with religion; modern science has fleshed out many of the details of *how* complex order is assembled and confirmed the overall Darwinian picture in impressive detail. The success of evolution, together with modern physics, is a leading reason for the dominance of naturalistic approaches in science today.

Nonbelievers who claim scientific support for a godless view of the world celebrate Darwinian evolution when arguing their case. They have good reason. The ancient argument from design—always the most popular reason to suspect that supernatural forces ordered our world—is no longer persuasive in a Darwinian universe. But even beyond the naturalism of evolutionary theory, there is something about evolution that is especially troublesome for religious perceptions of the world. Modern naturalism takes its framework from physics, and many of the central ideas in physics and cosmology can be religiously uncomfortable. But physics is rarely perceived as a spiritual nuisance, while evolution is regularly considered religiously dangerous. Nonbelievers often emphasize historical conflicts between science and religion, suggesting that these are two ways of thinking and two social institutions that are never compatible. And although focusing attention on conflicts can distort history, it is hard to deny that evolution remains the most famous source of friction between science and religion.

One reason is that evolution does, in fact, directly challenge commonly held religious views. Evolution says a lot more about the place of humans in the universe than, say, modern physical ideas about microscopic randomness. If there is anything at all to widely shared human intuitions about spiritual realities, it would seem life and creativity should be the responsibility of supernatural forces. Yet, according to Darwinian evolution, this is not so. And there is more. Not only does evolution motivate religious skepticism due to its uncompromising naturalism, but it also very easily leads into some traditional arguments for nonbelief.

For example, Darwinian variation and selection is in many ways a remarkably painful and inefficient process. Organisms usually produce offspring in numbers far larger than can survive to reproduce in their own turn, and selecting the most fit implies massive death and failure among the majority. Populations that strain the limits of resources, and the misery this implies, is the normal state of affairs in biology (Dawkins 1995). If we look at nature with anthropomorphic

eyes, its wonderful creativity also appears cruel, wasteful, and aim-
less. No intelligent designer working to achieve any purpose humanly
recognizable as good would do its work through natural selection.
Evolution, in other words, brings up the ancient problem of evil, the
most popular reason to doubt the gods. In fact, among nineteenth-
century evolutionary thinkers, including Darwin, the problem of evil
in nature was a significant reason to doubt the special creation of
species.

Nonbelief does not just draw on the philosophical tradition. The ab-
stractions philosophers argue over are usually too far removed from
actual religious life; most religious belief flourishes within particular
traditions with very specific doctrines and practices. Similarly, nonbelief commonly begins by doubts about a particular religious tradition rather than worries over the generic God of the philosophers. And
so evolution cannot help but get entangled with nonbelief, because it
is not compatible with traditional and fundamentalist versions of the
monotheistic religions.

The Darwinian picture of life is completely alien to the creation stories nurtured by biblical and Quranic literature, even when they are
not taken strictly literally. Whether creation took place thousands or
billions of years ago or whether a day to God is billions of years to us
are somewhat negotiable. However, beliefs that life was specially created and that humans were made in a divine image are religiously
much more significant. Like many religions, traditional Judaism,
Christianity, and Islam proclaim a *moral* order in a divinely created
universe. Creation stories do a lot more than satisfy curiosity about
human origins; they also serve as templates called on to justify a vision of social and personal morality.

So conservative religion has too much invested in special creation
to allow radical reinterpretations of sacred texts. In any case, evolution goes too much against the grain. Accepting that different life
forms are related is comparatively easy—monotheistic thinkers historically often placed everything on a "great chain of being" stretching from rocks to metals to plants to animals to humans to angelic
beings to God. The Platonic essences defining each species came from
God, in a top-down fashion, and humans existed midway between the
angels and the beasts. Darwinian evolution, however, is relentlessly
bottom-up. It has no more use for a chain of being than for the Genesis story of creation. It cannot be patched onto the traditional religious vision without doing violence to both.

This is not to say that sacred texts cannot be reinterpreted. Modernist religions, such as liberal Judaism and Christianity, or New Age religious movements expressing individualist and occult spirituality, typically affirm some version of evolution. They also promote much more modern attitudes about morality. However, liberal religious views of evolution, though affirming common descent, play down naturalistic mechanisms. Instead, theological liberals typically portray evolution as a progressive spiritual unfolding, an ascent up a ladder left over from the great chain of being. This halfway accommodation with science also affects religious practice. Liberals easily reinterpret the creation stories, but they too often end up treating the old stories as embarrassments to explain away, not as texts with a vital role in everyday life.

Reimagining creation stories, then, is both easier and more difficult in the modern world. It is easier because it has become hard to impose an official interpretation of texts on too many people at once. Modern individuals have become too aware of the existence of many spiritual traditions and too reluctant to accept institutional assertions of authority. So reading sacred texts in new and individual ways comes easily to many of us—it is just more religious creativity in changing times. What is hard is to make any interpretation stick. In the formative days of the Catholic Church, church leaders demanded that many scriptural texts should be interpreted allegorically. This was a common way to both avoid intellectual embarrassment among elites and to make already ancient texts speak to new circumstances. But such strategies cannot work today, because in the Western world there is no longer any authority that can coercively control interpretation. Modern people are still religious, but they also live in a world of a thousand heresies and actively engage in making more heresies all the time. It is easy to have faith that a text expresses divine wisdom, but very difficult to agree on what the text means.

For preserving both a sense of sacredness and a definite meaning in religious texts, the most viable option today is scriptural literalism. And this is an important reason behind the worldwide success of populist fundamentalisms. More sophisticated believers rightly point out that fundamentalist claims to be faithful to the traditional understanding of their scriptures is exaggerated; literalism is rarely the mainstream approach in religiously stable times. However, literalism is easy to assert and does not depend as strongly on central authorities to make sure the correct interpretation is followed. In today's populations that enjoy high literacy rates, literalism is easily understood and is readily available to all. Through fundamentalism, the common people can assert themselves against elites. So the more democratic, egalitarian political environment of modern times, which has allowed nonbelievers unprecedented social space, can also nurture religious fundamentalism.

The religions nonbelievers react against, then, are very often literalist movements that explicitly deny evolution. And so nonbelievers in strongly religious environments such as the United States and Muslim countries feel a need to defend evolution in what they see as a clash between science and religious unreason. Nonbelievers are concerned, and not just because they derive support for their nonbelief from the naturalism in science and evolution in particular. They have plenty of political motivation as well.

So the public debate over evolution and religion is perhaps more political than intellectual in nature. Even among secular people, evolution is often taken not just as a sensible account of our place in nature, but as a theory with immediate social implications. Left-wing secularists, for example, have often harbored suspicions of the Darwinian picture because of the role competition for resources plays in reproductive success and failure. Conservatives, in their turn, have often worried about variation and the possibly morally relativistic character of a Darwinian world.

Religion is about far more than explaining our world with the help of supernatural forces. Renouncing religion is also typically an act that involves more than accepting a naturalistic description of reality. The relationship between science and nonbelief has political complications, and this political aspect becomes especially visible in the debate over creation and evolution.

OLD-FASHIONED CREATIONISM

In political battles, opposing forces describe each other using negative stereotypes. So secular intellectuals regularly portray creationists as a narrowly sectarian minority, hostile to science, who want to take us back to a dark age. The reality is more complicated.

Wholesale denial of evolution is not restricted to a few minority sects—it has wide appeal to traditional and fundamentalist monotheist believers. This is a very large worldwide constituency, including some of the most successful religious movements of our time. Creationism sometimes seems an American phenomenon, since the United States harbors the most influential conservative religious population among industrialized countries. The most common "scientific" creationist arguments have an American pedigree. Nevertheless, creationism is even more widespread in the Muslim world. American Protestant creationism turns out to be easily adapted to an Islamic setting. Conservative Muslims have similar reasons to oppose evolution, as they also want to retain the sense that they live in a world that is a divine creation with a built-in moral order. And Islamic creationism is more successful: it often receives official sanction and has a strong popular appeal. Even the intellectual high culture in Muslim lands is cool toward Darwinian evolution; creationism finds favor in both popular and intellectual circles (Edis 2003b).

Creationists are also not antiscience in any straightforward sense. Creationism does not arise among peasants and traditional clergy; it appeals to modernizing populations who are fully aware of modern technology and who depend on technology for economic advancement. This is why creationists claim to be just as scientific as their evolutionist rivals, if not more so. Creationists are very positive toward technology (see Figure 3.2), and they recognize that science enjoys a reputation of correctly describing the world. That is exactly why they seek *scientific* validation for the beliefs that are vital to their communities. Since mainstream science cannot affirm their creation stories, they create an alternative "creation science" they claim is better science than evolution (Eve and Harrold 1991). Creationism is not just a stubborn assertion of faith—in its own way, it is an attempt to reconcile science and old-time religion.

Creationists also have no interest in another "dark age." They want science to confirm their religious views, but they do not ask to subordinate science to anything like a church. Fundamentalism is a modern way of being religious: populist, egalitarian, affirming of common sense—the kind of common sense that allows ordinary individuals to usually agree on the literal meaning of sacred texts. Creationist opposition to mainstream science has as much to do with scientists constituting an elite as with the religious difficulties inherent to evolution.

Nevertheless, creationists do deny evolution, and their differences with the mainstream scientific community are very real.

First and foremost, politics aside, creationism is an intellectual disaster. Scientists do not favor evolution because of a philosophical whim; the structure of evolutionary theory, the evidence in support

Figure 3.2 Creationists often are very enthusiastic about applied science. This cartoon is from the "After Eden" series appearing on the web site of Answers In Genesis, a creationist organization. (Courtesy Dan Lietha)

of it, how it fits in with the overall body of our scientific knowledge makes it near impossible for most scientists to deny evolution. Creationists call for a massive overhaul of science, but then present only very dubious arguments to persuade others that creation is the better scientific position. For example, creationists typically assert that the second law of thermodynamics makes evolution impossible. No self-respecting physicist would agree; yet, creationists do not retract or significantly revise their second law argument. Indeed, creationists produce no end of appallingly bad scientific claims in their publications. Protestant creationists attempt to account for geology by Noah's flood, promote strange cosmologies that would restore Earth to the center of the universe, refuse to acknowledge transitional forms in the fossil record, and so on (Morris 1985). Scientists find all this very hard to take seriously.

While creationists may be enthusiastic about technology, they have an impoverished conception of science. They distrust theory, accepting applied science or concrete experimental results but brushing aside the conceptual schemes of modern science. They take a commonsense view: there are facts, and then there are theories that are little better than guesses. Naturally, scientists are not likely to adopt a view of science that would deny the importance of theory and thereby reduce science to stamp collecting. Scientists care about their accomplishments—they think they have a good, though not perfect, handle on how the world works. In countries such as the United States, it is no surprise that scientists are seriously concerned about a popular movement that does not just oppose biological evolution but would require radical surgery on the whole modern scientific enterprise.

Intellectual disagreement alone could be contained. Scientists might consider creationism foolish, and creationists might think evolutionists are bound for hell, but they could live together. Unfortunately, they compete for resources, and so the political battle heats up.

Education is the biggest battleground. Creationists do not like their children being taught ideas that may undermine their faith. Science has some authority in our culture, so what is presented as science is understood to be factually true. Requiring the teaching of evolution in public schools is governmental sponsorship of an idea that is offensive to conservative religion. Creationists would prefer that schools aid in reproducing their culture, or at least that their taxes should not be used to oppose their beliefs.

Scientists have an interest in education as well. Many scientists are partly supported by their teaching in schools of higher education, and the new generation of scientists is recruited from students who already must have had some scientific and mathematical training before they reach college. Moreover, science education is an important way by which scientists hope to influence the wider culture to respect science and provide resources for scientific enterprises.

To further complicate matters, other secular constituencies, including most nonbelievers, make their own demands on mass education. Schools, they hope, can further secular pluralism. If exposure to modern science nudges students away from belief that any one religious tradition has exclusive truth, this can only be for the good.

Debates over evolution, then, are rarely purely intellectual affairs. They involve politics, struggles between conservative and liberal religiosity, and worries about public morality far removed from biologists' thoughts about evolutionary relationships between species of snails. In this battle, scientists and nonbelievers have a common interest in combating the public influence of creationism.

INTELLIGENT DESIGN

Old-fashioned creationism has a solid constituency among religious conservatives, but it is an intellectual embarrassment. So educational and scientific elites rarely take creationism seriously except as a threat. For nonbelievers who suspect piety and anti-intellectualism go hand in hand, creationism is a powerful confirming example.

Conservative religious thinkers have therefore felt a need for a more sophisticated defense of creation. The recent intelligent design (ID) movement attempts to present an academically respectable opposition to evolution. ID proponents shy away from explicitly religious appeals, showcasing Ph.D.'s who teach in mainstream universities. They avoid arguments based on scripture, and steer away from absurdities such as claiming that Noah's flood was responsible for the geology of the earth. Instead, they claim that biological structures are too complex for Darwinian mechanisms to assemble. They think some outside intelligence must have intervened. More abstractly, ID proponents argue that *information*, such as that in DNA sequences, cannot be generated through chance and necessity. That something more than blind nature is responsible for life is a common conviction, especially in theistic religions. ID tries to flesh out such basic religious intuitions about design and complexity in scientific terms, and then show that life is, indeed, designed (Dembski and Kushiner 2001).

Given how entrenched the Darwinian view of life is in mainstream science, ID does not face an easy task. Naturalistic approaches such as evolution have been so successful that most scientists today would not look outside nature to explain anything they study. So ID has to start by bringing design back into science as a legitimate option.

The difficulty of this is that ID would then threaten the accommodation between science and liberal theology. Science, according to a common liberal theological view, is limited to finding natural causes for natural events. While science cannot, for example, invoke a creator-God in its theories, it has no business suggesting that no creator exists either. Whether a scientific theory is true (rather than just useful), what it means, and how it relates to God are questions for philosophers and theologians, not scientists. Politically, limiting science to naturalistic explanations serves both science and liberal religion. Scientists do not have to worry about religious interference, and religion is protected from science-inspired criticism.

So in the United States, scientists and religious liberals typically resist old-fashioned creationism by defining science as a search for natural explanations. Public defenses of evolution often state that one can easily be religious and believe in evolution, and that creationists make a basic mistake in trying to introduce the supernatural into science. Science has its rules, and sticking to natural explanations is one of the rules that define science. Introducing a creator is like insisting on playing basketball in a bowling alley. In one quick stroke, defenders of evolution exclude creation from science, without having to say anything about whether evolution is true or what evidence supports it.

Mainstream science has very often responded to ID in the same way: by treating it as a philosophical mistake. If science operates according to "methodological naturalism," an intelligent designer outside of nature cannot be debated within science (Pennock 1999). Again, this approach is politically appealing. Unfortunately, it is intellectually weak. ID has attracted a number of academically qualified defenders, especially among philosophers and theologians. They have been quick to point out that *defining* science so as to exclude ID is not legitimate. If science is supposed to be our best shot at figuring out how the world works, and if we exclude an intelligent designer, this is prejudging how scientific inquiry should turn out. And even if we were to accept a narrow and arbitrary definition of science that excludes certain explanations, the question of how complexity arises will not go away. To properly address that question, we would need to broaden our perspective, and weigh the merits of the hypothesis of an intelligent designer in a way much like how we deal with naturalistic proposals. After all, on the face of it, the claim that biological information did not evolve but was intelligently designed is a perfectly intelligible proposition, whether it is true or false. Mainstream science has little difficulty accepting intelligent design by humans as a legitimate explanation when studying ancient artifacts, or even when attempting to detect radio signals sent by extraterrestrial intelligences (Dembski 2004). The claim of an intelligence beyond nature is more ambitious, but it is not beyond investigation.

Much of this is correct. Excluding the very possibility of ID from science would be arbitrary, and would have more to do with politics than science. Those philosophers of science who have sought a defining essence of science have not been very successful, and limiting science to the purely natural fares no better. So the philosophical concerns of ID proponents have some merit. The task of ID, however, is not over. Making methodological naturalism a rigid requirement for science is a bad idea, but conceding this point only lets ID into the court of science. It still may be judged an unsuccessful explanation of the world. While not a rigid defining principle, methodological naturalism is a sound practice for scientists today. After all, scientists want to devote their efforts to pursuing explanations that have a good prospect of success. For example, few scientists would jump to a conclusion that some peculiar sightings of lights in the sky are due to alien spaceships, especially if any possibly interesting signal is lost in a noise of hoaxes, perceptual mistakes, and cultural myths in the making. Similarly, naturalistic ideas enjoy the best prospects in science today—since the scientific revolution, supernatural explanations have steadily been replaced by much more successful naturalistic alternatives. The presumption that naturalism will continue to succeed can be challenged, but not without solid scientific work in favor of a claim like ID.

ID proponents think they have a good scientific case. In biology, ID rests on the work of Michael Behe, who argues that certain molecular machines and biochemical processes in cells are "irreducibly complex" (1996). To take the favorite ID example, some bacterial flagella look like an outboard motor. A number of well-matched parts have to come together simultaneously for the motor to function: removing any one makes the flagellum fail completely. Behe argues that such a system is too complicated to be gradually assembled through variation and selection. After all, any intermediate steps between having a flagellum and having none must themselves be advantageous to a living bacterium. But intermediates must have missing parts, leaving the bacterium with a worse-than-useless appendage until everything falls in place.

Behe's argument, however, is not new. It updates a creationist classic that asks what good would half a wing be for an animal. The answer is, better than no wing at all, but not for the purpose of flying. Biologists call this *exaptation*: anatomical features that serve one function in intermediate stages can be diverted to entirely different uses in the course of evolution. Protowings in the dinosaur lineage that led to birds, for example, can start out as useful tools for trapping insects and then gliding before full flight becomes possible (Gishlick 2004). The story is no different at the molecular level. Bacterial flagella are not simply outboard motors; they are in fact secretory structures as well. Though knowledge of the details is sparse—flagellum evolution does not leave historical records such as fossils—biologists working in this area consider it very likely that flagella developed from secretory mechanisms (Musgrave 2004). Earlier versions of flagella did not, in other words, always perform their present function. They found a use in moving bacteria around after they had already evolved for different reasons.

Mainstream biologists have not ruled out Behe's argument because it was insufficiently naturalistic; they have considered it and decided it has no merit. "Irreducible complexity" does not trouble Darwinian evolution.

William Dembski, however, presents a deeper challenge to evolution (1999). Intelligent design (ID) is concerned with biology mainly because life is the best example of the sort of functional complexity ID proponents think signifies design. Its real focus is intelligence and information. Dembski attacks naturalism head-on. Intelligence, he says, is an entirely independent principle, not reducible to chance and necessity. He also claims to have developed a mathematically rigorous and reliable way to detect "complex specified information" (CSI), which cannot be produced through chance and necessity. CSI is supposed to be a signature of intelligent design, and organisms are supposed to be full of CSI.

Dembski's apparatus to detect design does not work. His CSI is based on a badly defined probability estimation. Complexity and information are well-explored concepts in physics, information theory, and computer science, and Dembski's work makes no real contribution to these fields (Perakh 2004; Young and Edis 2004). Furthermore, there is no real hope to patch up Dembski's notion of CSI and discover a sort of information that signifies design. Scientists already know how combinations of chance and necessity—Darwinian evolution in particular—generate new information. It is especially important that evolution is an open-ended process, with no preset target. Because of random variation, evolution can explore genuinely new options (see Figure 3.3). And since evolution affects the very environment to which organisms are adapting—other members of the same species plus competing forms of life are often the most important aspects of an individual's environment—just what evolution will produce is not pre-

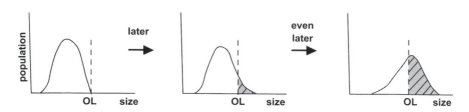

Figure 3.3 Random variation introduces novelty into the genome. Here, the size of individuals in a population changes over time, evolving to extend beyond the original limit (OL) in sizes. (Ricochet Productions)

dictable. Everything is subject to variation, and so the information generated through Darwinian evolution is genuinely new. It has no source outside of nature at all. In fact, there is good reason to believe chance and necessity can account for all we see in biology, including the results of intelligent design by humans. Far from being a separate principle, intelligence itself is most likely a product of variation and selection-based processes in the human brain (Edis 2004b).

Again, ID is an intellectual failure. It fails not because the supernatural is out of bounds to science, but because when judged scientifically, ID has nothing real to offer. On the philosophical side, ID has some minor substance, as it rightly points out that the conventional argument for excluding creation from science is not neutral toward religion. Ruling the very claim of ID inadmissible adopts the perspective of a liberal religiosity that shies away from letting supernatural forces act overtly in the world. And requiring that evolution should be taught to students, with the implicit understanding that science gives us the best picture of how things really are, favors modernist over conservative religions. Given the weakness of philosophical arguments against ID and the political conservatism of the United States, the ID movement can perhaps hope to affect U.S. educational policy more than old-fashioned creationists. But its ambitions to bring about a scientific revolution will get nowhere; ID simply has nothing to offer to science.

So the debate over ID ends up in the political arena, and ID appeals to the same constituency as old-fashioned creationism (Forrest and Gross 2004). Supporters try to force ID into American high school science classes, even though ID-driven research is virtually nonexistent in the scientific literature. For nonbelievers, however, the politics of opposing ID is more complicated than with creationism. ID is a concentrated intellectual effort to remove creativity from the natural world, to restore a clear top-down sense of design to nature. Its scientific failure can only reinforce the conviction of nonbelievers that our world is not a divine creation. Nevertheless, the philosophical aspect of ID does identify a weakness in the arguments that have allowed secular people to join with religious liberals to keep creationism out of education. Answering ID claims is best done by pointing out how scientifically worthless they are. But admitting that supernatural claims about creation can be vulnerable to scientific criticism is politically dangerous.

CREATION THROUGH EVOLUTION

Nonbelievers and religious liberals are allied against creationism, but they differ in their views of evolution and its implications. The issue is not whether a religious believer can accept evolution—many do, obviously. Devoutly religious evolutionary biologists are not an endangered species. Also, evolution is clearly compatible with the existence of a God in a trivial sense: religious people who accept science do not endorse a logical contradiction. Even so, Darwinian evolution is vital for modern naturalism, while the religious have to struggle to see how Darwin's vision can be fitted into a spiritual worldview. Evolution seems to *count against* spiritual realities.

So religious thinkers have tried to incorporate evolution into their beliefs about creation. The most popular approach is to interpret evolution as a kind of progressive spiritual unfolding. There once was no life or mind or anything spiritually significant. In time, life started and became more complex, until conscious intelligent creatures came into being. This history, many have thought, indicates that the universe is on a path of spiritual progress, and that evolution is God's instrument of creation. God directs evolution, ensuring it will produce human beings fit for a spiritual relationship. Catholics say a soul is infused into people after evolution constructs suitable bodies. Academic theologians speak of how divine love draws recalcitrant matter toward itself (Haught 2000, 2003).

Progressive evolution is certainly popular. If about 45% of the U.S. population has creationist sympathies, an equal number think evolution is divinely guided (Bishop 2000). We tend to think of evolution as ascent up a ladder and praise not only species but individuals as being more "highly evolved." Liberal religions and New Age spiritualities often not only accept evolution, but portray the whole history of the universe as a cosmic progression of great religious significance.

The trouble with progressive evolution is that it has little to do with actual biology. The mechanisms of evolution are completely blind. There is no inherent direction, progressive or not, in evolution. There are trends: some life forms have diversified into increasingly more complex ways of living over time. These trends, however, are fully explained by blind mechanisms and the fact that the starting point was one of simplicity. There is no extra outside force nudging evolution in a progressive direction. Bringing in divine guidance would be like using conventional physics to explain how a cold object and a hot object will equalize in temperature when brought into contact, and then, on top of that, also proposing an occult force that purposefully equalizes temperatures.

For progressive evolution to work, genetic variation must not be blind—it must favor spiritually relevant developments. Some scientist-theologians, for example, think non-Newtonian physics allows covert divine influence. God can tweak evolution, ensuring that something like humans turn up; perhaps by injecting "active information" under cover of physical randomness (Polkinghorne 1998). Other liberal religious thinkers also speak of God being the ultimate source of the information manifested by evolution. Like intelligent design, liberal versions of guided evolution also express the intuition that information must have been injected into an otherwise inert material world by an outside spiritual force (Dembski and Ruse 2004). Liberal ideas, however, are much more speculative; unlike ID, they do not overtly challenge mainstream science. They suggest that what most scientists think is blind is really not so, but do not mount any scientific defense of their claim.

A more promising approach is to accept that evolution is blind, but argue that it was still a device set up to achieve humans. After all, the operation of a television set is fully described in mechanical and electrical terms. Still, the physical explanation is incomplete, because it leaves out the purpose that brings the physical parts of the TV together in exactly the right arrangement. Similarly, evolution may work through blind mechanisms, but may still be a device to achieve a divine purpose.

If it serves a purpose, evolution must produce reliable results. The broad outcomes of evolution are roughly predictable: starting from simple replicators, if variation and selection can explore an increasing variety of ways of life, *some* replicators will become more complex. Increasing brainpower in some animal lineages, for example, is no surprise. It is still not clear, however, whether conscious, highly intelligent animals such as humans were inevitable. There is no scientific consensus on an answer. Stephen Jay Gould, for example, plays up the historical contingency of evolution, and says that if we could return to the conditions of a few hundred million years ago, we could not guarantee something similar to humans would ever turn up again (1989, 1996). Simon Conway Morris, in contrast, emphasizes convergent evolution. Certain good solutions to common problems will emerge again and again, in the way many animal lineages independently evolved wings and flight. Morris thinks human intelligence and the human body form are nearly inevitable; he goes on to argue that evolution could be God's way of creating us (2003).

Adding the supernatural on top of evolution is not, however, so easy. If bringing in the gods gave us an independent way of understanding the details of evolutionary history, as with the TV and its designers, we would have a fuller explanation. After all, for pure naturalists, many of the twists and turns of the history of life are sheer accidents. Things could have turned out differently; for example, mammals would not have achieved prominence if the dinosaurs had not been wiped out by a catastrophe. Naturalists have to leave such details unexplained, saying things just happened to work out that way. So a better explanation of those details can supplement or overturn naturalism. Just looking at the results of evolution and declaring that such may well have been the divine intent all along, however, is not an explanation.

Imagine a tornado ripping through a town, ripping apart Smith's house but leaving her neighbor Jones's intact. A naturalist could not give any deep explanation for why Smith was hit and not Jones. Ultimately, it was blind luck. A supernaturalist, though, might say the tornado's course was set by a god who punished Smith's sins; or, for that matter, by a demon who decided which house to level based on the color of the owner's underpants. Strictly speaking, a tornado might be compatible with supernatural purposes beyond the physics, however absurd. Still, the supernatural element adds nothing to an explanation of the tornado. It removes an accident from the natural world, only to replace it with arbitrary characteristics of a supernatural reality.

Liberal theological interpretations of evolution try to avoid doing violence to the science. But such interpretations look like adding a divine purpose to a tornado. Compatibility with science comes at the cost of constructing a cosmic Santa Claus that explains nothing. Old-fashioned creationists get it wrong, but liberals, nonbelievers suspect, have a way of not even being wrong.

Still, hopes to find a more comfortable fit between evolution and the gods are not likely to vanish soon, as long as science and religion both remain culturally important. Any new development in science is likely to be harnessed to this end, in however half-baked a fashion. Today, words like "self-organization" and "emergence" detach from scientific use and float down into the wider culture as buzzwords. Just as New Agers use "quantum" to stand for magic, "emergence" has lately become a favorite word in arguments that evolution was somehow ordained by a God. The fecund capacity for self-organization, complexity, and evolution of the universe is supposed to be due to a supreme intelligence (Gregersen 2003).

Theologians of the "emergent" version of progressive evolution start by saying the whole is more than the sum of its parts. Indeed, it is practically impossible to understand the behavior of strongly interacting complex systems by attempting to compute the interactions of its components in detail. No one can hope to understand a bacterium by starting at the level of particle physics. Scientists instead identify structures such as membranes or genetic material that have their own identity and integrity, and work at this level. However, scientists also try to see how these higher-order structures emerge from the more basic physics, without inventing any new particles or interactions to do the trick. Self-organization is about understanding how parts make up wholes; it is the opposite of "holistic" mystification. In particular, nothing in today's physics or biology supports the notion that once we identify a self-organized whole, we can also see an independent top-down sort of causality by which higher levels act on and even construct the lower levels. All the real explanatory work in self-organization is done bottom-up.

Theologians of emergence also suggest that "higher" levels of organization are better—more capable, more intelligent, and so forth. However, such humanlike qualities we admire are adaptations to our particular corner of the universe; there is no reason to identify them with some alleged emergent progress. The physics of complexity gives us some general principles independent of the microscopic physical substrate from which a complex system emerges. But these principles are such things as thermodynamics or Darwinian evolution; they have no hint of anything so anthropomorphic as spiritual progress. We can no more expect intelligence to generalize across levels of organization than we can expect the spin of elementary particles to resurface in our descriptions of macroscopic objects. Anthropomorphizing emergence ignores how profoundly *different* levels of description generally are.

Such misrepresentations of science as progressive evolution are less of an irritation to the scientific community than a politically powerful creationism. This may change, however, as New Age spirituality becomes more influential. Beliefs like progressive, emergent spiritual development are woven into the supernatural ideas supporting alternative medicine. As New Age spiritual healing continues to penetrate the medical fields, which is a huge source of funding for biological research, scientists will likely become more concerned. The amicable relationship between liberal religion and mainstream science could then come under strain.

UNIVERSAL DARWINISM

Scientists rarely object to interpreting evolution in a way compatible with a God, as long as the interpretation is purely philosophical or theological. Such interpretation comes as an optional add-on, making no demands on the task of scientific explanation. Religious conservatives dislike this option because it makes the supernatural redundant. Nonbelievers think relegating God to the inconsequential talk of theologians does not go far enough.

So nonbelievers inspired by science keep coming back to how evolution needs no gods and how it solves the problem of creativity and complexity entirely within the natural world. Biologist Richard Dawkins famously points out that "Darwin made it possible to be an intellectually fulfilled atheist" (1986, 6). Both for his public criticisms of religion, and for his popular expositions of Darwinian evolution, he has become a favorite among nonbelievers. Indeed, Dawkins repeatedly points out how Darwinian evolution, with its blind mechanisms that take no account of morality or suffering, is just the sort of creative process one would expect from a world that is purely natural, indifferent, and mindless at bottom (1995).

Arguments like those Dawkins makes support worries that evolution has overstepped its proper bounds as a scientific theory and become a godless worldview. Those sympathetic toward creationism or ID even suspect that Darwinian evolution was thought up precisely as an atheistic philosophy, to fuel the rebellion against traditional religion brewing in nineteenth century Europe. The creationists are wrong. However much Darwinian ideas have been congenial to nonbelievers, variation and selection was a deeply original idea, going far beyond the well-worn themes of the philosophical tradition of nonbelief (Bowler 1988). Nevertheless, religious thinkers are correct to observe that Darwinian thinking has overflowed its biological bounds. Today, we find Darwinian explanations not just in biological fields such as medicine, but also in psychology, brain science, even artificial intelligence and physics. Wherever adaptation, creativity, and complexity beyond rigid physical constraints occur, the naturalistic explanation involves a component of variation and selection. A "universal Darwinism" is taking shape.

So today, the Darwinian impetus toward nonbelief is stronger than ever. Daniel C. Dennett (1995) describes the Darwinian idea as a "universal acid" that eats away at all supernatural beliefs, at any top-down conception of reality. Increasingly, impotent "interpretation," which removes the divine from any substantial contact with reality, seems to be all that theologians can do. Even if not strictly atheistic, the acid of universal Darwinism is integral to modern naturalism, and is truly corrosive of religious belief. Many a religious student who encounters evolution in class begins to question the creation stories they grew up with. Creationists are not being paranoid when they see evolution as a threat to their religious heritage.

Moreover, even some of the peripheral conservative worries about evolution have some justification. A very common theme in antievolutionary thought is that when a society accepts an evolutionary account of human origins, social corruption follows. Charges that belief in evolution unavoidably leads to communism, Nazism, racism, and so forth are absurd, but evolution does, in fact, undermine a common traditional conception of the nature of morality. In a Darwinian world, nature is no longer infused with morality. Living things do not have created functions that are right and proper, and variation is not a deviation from an essence with overtones of corruption. For example, homosexuality cannot, in a Darwinian world, be condemned as "unnatural" or suspect because it deviates from the created sexual function. In today's debates over homosexuality, liberals and conservatives hotly dispute whether homosexuality is rooted in genetics or lifestyle choices—implicitly equating the genetic or natural with the morally acceptable. Thoroughgoing Darwinians, on the other hand, might end up judging homosexual conduct to be either worthy or socially undesirable, but their moral thinking would no longer be anchored to the traditional conceptions where right and wrong are reflected in created nature.

Evolution, then, inevitably figures in questions of morality, politics, and religion, whether the scientific community likes it or not. Darwinian evolution pulls people toward naturalism, but being associated with religious nonbelief is, in most parts of the world, a political liability. So while nonbelievers are perhaps the most enthusiastic supporters of evolution aside from scientists themselves, scientists also have an interest in distancing themselves from public expressions of nonbelief.

Their political weakness puts nonbelievers in a curious position when promoting evolution. For example, in the United States, there is a recent movement to celebrate February 12, Darwin's birthday, as "Darwin Day." This event is supported largely by humanist, freethought, and atheist-oriented groups, using slogans such as "a celebration of science and humanity." Naturally, the scientific community responds positively, treating it as a public outreach. A biology professor, for example, might give a talk on evolution to a secularist student group as part of their Darwin Day event. Occasionally, university science departments cosponsor larger public events put on for Darwin Day, alongside atheist and humanist organizations. Their relationship is, however, awkward. Scientists do not like their work to be perceived as political or ideological in nature, and nonbelief is one of the politically least desirable constituencies with which to be associated. Public presentations in Darwin Day events include explorations of Darwin as a freethinker and talks that highlight evolution's challenge to religion. But by and large, scientists especially remain comfortable repeating the conventional wisdom, which is that evolution threatens only a small minority of fundamentalist believers, and that liberal theological views of science and religion are indisputably rational and correct (see Chesworth et al. 2002). Nonbelievers, in other words, end up supporting liberal religion.

An alliance with religious liberals need not bother the nonreligious. After all, nonbelievers most often react against politically intrusive, conservative religions. Their political goals and ethical inclinations are usually close to those affirmed by modernist spiritualities. And even those nonbelievers who equate all religion with superstition very often think religious liberals are already halfway to rejecting the gods. If so, promoting public acceptance of Darwin would also nudge people toward dropping their supernatural beliefs, even if they hang on for a while to vague liberal conceptions of divinity.

The politics of evolution is, however, considerably more complicated. Liberal religion is not a mere compromise between old-fashioned faith and newfangled rationalism. Liberals resemble nonbelievers in their nonauthoritarian social ethics, but they still try to anchor their sense of meaning and morality in something transcending nature. Science may be successful in its own domain of the material world and even socially useful. However, liberals also want to limit science, to prevent it from shaping the world of meaning. Though natural science has been the most spectacularly successful intellectual enterprise of the past few centuries, it is not the only form of scholarship. Theological liberals often draw on ways of thinking common in the humanities to suggest that science cannot address the really important questions. Some, especially those attracted by postmodern philosophy, reject any claim that modern science has the kind of hold on reality that should compel anyone to adjust their view of the world. Others are more concerned to reinterpret religion and find some way in which claims that go beyond nature appear to be compatible with science. For if science is limited and compatible with religion, religion can be validated by its own means, entirely independent of scientific matters.

In fact, believers of a more fundamentalist stripe can be, at a deeper level, more positive toward science than liberals. They accept that science is compelling, and they are not convinced science can be confined to a religiously irrelevant domain. This is why they so often insist that science—what they consider to be science done properly, without the materialistic prejudices of mainstream science—fully confirms a religiously useful picture of the world. Many fundamentalists and nonbelievers have similar conceptions of science and religion; they agree that science represents our best shot at figuring out the nature of our world, and that any worthwhile supernatural reality should leave its stamp on the world in a way science can recognize. Their debates often focus on claims of fact rather than meaning. And so fundamentalists and nonbelievers alike are regularly accused by religious liberals of having a superficial conception of spirituality.

However strongly Darwinian evolution weighs in on the side of naturalism, nonbelievers cannot expect that acceptance of evolution will translate to a decline in supernatural belief. And no group of people concerned with the debate over evolution—scientists, nonbelievers, and religious people of all kinds—can count on the political landscape remaining stable. Evolution will continue to produce friction between science and religion, both intellectually and politically, for a long time to come.

Chapter 4

Minds without Souls

MIND AND MATTER

Physics, chemistry, and biology describe their objects of study in strictly naturalistic terms. They challenge spiritual perceptions of the world, presenting us with a universe where life and creativity are not handed down from above but bubble up from below. The ambitions of modern science do not, however, stop here. In the last century or so, scientists have begun to explain the phenomena of our minds—emotions, intelligence, consciousness, and all—in the same way as they have linked up biology with physics.

In fact, understanding minds as a physical process and no more has come to seem inevitable. After all, the world of physics appears closed, allowing no outside ghostly influences. And Darwinian evolution means that humans are animals. A special sort of animal, no doubt, but still an animal, with impressive mental capabilities assembled by evolution just like our bipedal posture or opposable thumbs.

As Owen Flanagan points out (2002), however, the emerging "scientific image" of persons clashes with a more familiar "humanistic image." According to the accumulated wisdom of our culture, people are fundamentally spiritual beings. Our minds have a nonphysical, immortal component. We are created in the image of God, and being so created, we possess free will—a godlike power of acting on the world because of our own decisions, unconstrained by anything outside ourselves. In fact, it is just because humans are partly outside of mere physical nature that we can have moral aspirations and find meaning in life. If we are, as the developing scientific image has it, fully physical beings, human dignity comes under threat. There are no spirits or souls; when we die, we no longer exist. Whatever free will we possess cannot be of the godlike variety. We have to negotiate moral agreements rather than discover absolute moral truths. People with strong religious sensibilities will fight such claims tooth and nail. But the scientific image can seem threatening to nonbelievers as well. Many people discard religious beliefs and adopt a secular way of life, but still hold to prescientific beliefs about minds. If materialist conceptions of mind are on the right track, many secular moral philosophies must be deeply flawed, no less than traditional religious doctrines.

So the scientific image of a person seems dangerous. Moreover, outside of a small circle of scientists and philosophers impressed with science, the scientific image usually appears manifestly absurd. A collection of physical particles can have all sorts of interesting properties, such as an electric charge. A physical process can accomplish wonderful things, like a computer that impressively though uncomprehendingly solves a difficult set of equations. But minds, it seems obvious, are not about physical properties at all: they involve thinking, meaning, awareness, believing, fearing. How can a set of physical interactions, even those making up such an immensely complex object as a human brain, ever add up to such mental goings-on?

Modern science is accustomed to violating deep intuitions and everyday common sense. So if the scientific image of the mind goes against our ingrained folk psychology, this is no great surprise. Still, folk psychology is not easily trifled with. It is perhaps understandable that our folk-physical intuitions are limited, and that folk biology fails to anticipate evolution. Psychology, though, is something we are good at. Folk psychology is immensely rich and very effective when negotiating the world of our fellow human beings. Much of human thinking is devoted to figuring out the wrinkles of social interactions: most of us are remarkably good at reading emotions, forming elaborate accounts of people's intentions, and detecting possible cheating in social circumstances. Observing certain mental patients, autistics, and other people impaired in their folk-psychological abilities makes it clear how vital intuitive psychology is for our lives. The best instrument we have for taking the measure of a normal human mind is still a well-trained, empathetic, observant person who understands her subjects in commonsense terms.

Belief in a soul, or at least the belief that minds are radically different than matter, is deeply ingrained in folk psychology. Just as our brains automatically distinguish living from inert objects and generate different sets of expectations about how they should behave, we treat personal agents as a distinct category of entities in the world. Likewise, we treat agent causation—involving intent and purpose—differently than mindless physical causation. Even everyday evidence for souls is not hard to come by. For example, we all dream. The most natural explanation is that something spiritual leaves the body to have conscious experiences unfettered by earthly constraints. If we dream of meeting other spirits, even spirits of the dead, maybe this is just the sort of thing a disembodied soul would experience.

So it would seem that even if the supernatural is not easily found in physics or biology, religion need only look within. For our minds, our very selves look like they are nonphysical. If so, the spiritual is as intimate as our selves. Supernatural agents with power over physical reality make as much sense as souls in control of bodies, affecting the material world as directly as we raise an arm by an act of will.

Scientists who want to realize the grandest ambitions of physics and nonbelievers inspired by natural science hope to dethrone the humanistic image of persons, however deeply ingrained. Amazingly, their image of persons has become the more confident view in intellectual life today.

CARTESIAN DUALISM

Though our intuitive psychology suggests that there is a spirit realm, the claim of souls needs a clearer statement if it is to be intellectually respectable. Until recently, one of the most influential theses about mind was Cartesian dualism. The seventeenth-century philosopher, mathematician, and scientist René Descartes argued that mind and matter were radically different from one another. Material objects were governed by the iron laws of physics. Mind, however, was a thinking substance lacking all physical properties such as extension in space.

Cartesian dualism fit traditional religions very well, declaring that we are, in essence, immortal souls. Less obvious today is how dualism was also useful for science. By the way they imagined the mind as a supernatural substance, Cartesians allowed investigators to see the material world in completely mechanistic terms. Outside of human minds, everything in nature, including animals, became manipulable. Scientists could isolate a few variables, do controlled experiments, and hope to get to the heart of the matter. This approach contrasted with the way the occult tradition suggested investigation should work. An alchemist or astrologer would assume that life and spirit pervaded the universe, and so understanding the world required a spiritual journey (Burton and Grandy 2004). Rather than disassembling a machine to see how the parts work, the occult investigator would try to get in tune with the harmonies of nature, to achieve the right relationship with a living reality. Cartesian dualism emptied most of the world of spirit, allowing scientists to have an impersonal relationship to their objects of study.

Until about a century ago, dualism remained dominant. Not only was dualism the most common flavor in philosophies of mind, but much of science took it for granted. When the new science of psychology began in the nineteenth century, it intended to study the inner life. Psychologists relied on techniques of introspection, asking their subjects to produce elaborate reports on the contents of conscious experience. As physicists revolutionized their science in the early twentieth century, they assumed observations were available without analyzing observation as a physical process. Somehow, the world impressed itself upon the soul.

Dualism, however, ran into difficulties. As a method, introspection soon became a dead-end for psychologists. And philosophers kept being bothered by the problem of how a nonphysical substance could interact with the material world. Dualists have to assume that by some mysterious supernatural means, the soul enacts its will on the body and perception happens to inform the soul about the world. But the interaction is clearly tight. Material events can drastically affect the mind. Brain injuries, for example, can cause huge differences in conscious experience, as when a patient is completely paralyzed on one side of her body but does not realize this and continually denies any such paralysis exists. Changes in the brain can lead to radical personality changes and the kind of gradual deterioration seen in Alzheimer's sufferers. Drugs, notoriously, alter the mind.

Consider a radio receiver. Physically interfering with the radio changes or destroys the sound it puts out. But the mechanical structure of the radio alone does not account for the music. All the interesting information is encoded in the radio waves that the receiver translates into the mechanical waves we hear as sound. When radio waves were newly discovered and mysterious and seemed to possess occult qualities, some dualists made an analogy to radios; the brain was necessary to express the soul but the essence of mind was beyond physics. The analogy is not, however, persuasive. After all, radio waves are physical, and they physically interact with the receiver. Moreover, the relationship between brain and mind is too strong; it is difficult to say, watching a patient whose brain damage alters her personality or conscious experience, that her soul lives on unchanged, impaired only in its ability to express itself materially. Dualists need some account of how the brain and the soul interact, and Cartesian dualism, which insists that minds are utterly nonphysical, cannot provide such an account.

Moving beyond dualism, many psychologists first embraced behaviorism. Psychology was to be the study and explanation of human behavior, nothing more, and therefore behavior was not to be linked up with consciousness or beliefs or anything in the conventional vocabulary concerning the mind. People had behavioral dispositions and they learned by varieties of conditioning. Behaviorists insisted that after all, a proper science should only be concerned about observable behavior rather than invisible mental events. Philosophers also moved away from dualism, beginning to think that talk of a Cartesian "ghost in the machine" was a mistake of language (Lyons 2001, see chapters 1 and 2). In a time when many intellectuals had lost confidence in supernatural claims, behaviorism was also attractive because it promised to do away with spirits and other spooks. Many a nonbeliever was an enthusiastic behaviorist, and vice versa.

Behaviorism in psychology and philosophy did move science forward—for a while. Its limitations soon became clear. Behavioral dispositions were all very fine, but behaviorism could not give any deeper account of dispositions. In addition to which, it blatantly ignored our inner life. So thinkers about the mind again focused attention on what went on inside the head.

For those not willing to go back to dualism, the most interesting mid-twentieth-century development in philosophy was the attempt to identify mental states with brain states. A state of "believing that grass is green" was supposed to be literally identical to some sort of brain state, maybe that of a cluster of neurons firing in a characteristic pattern. Now, this identity sounds very implausible, but conceivably this is because we do not know enough about the brain. And at midcentury, the brain was undiscovered territory. So the mind-brain identity theory usefully directed attention to brain research. The trouble was, nothing coming out of the neuroscience of the time helped the identity theory gain any plausibility. Plus, narrowly identifying mental states with brain states was not very persuasive when actual human brains varied greatly in detail. If Smith's belief that grass was green was identical to what a certain cluster of his neurons were doing, what about Jones's same belief about the color of grass, when Jones did not have anything like the same set of neurons doing exactly the same thing? Philosophers identifying the mind with the brain had no clear idea *how* brain states realized mental states, and no useful suggestions for psychologists and neuroscientists to test.

Today, the philosophy of mind has a wide range of ideas in play. Varieties of dualism survive, especially among religious philosophers (Habermas and Moreland 1998; Hasker 1999). Others are functionalists inspired by computers, or eliminative materialists who think that as neuroscience matures the concepts of folk psychology will fade away. There are philosophers who call themselves nonreductive materialists, those who believe the mind-body problem is unsolvable, and more. Indeed, the fact that philosophers are still so intimately involved in theorizing about the mind is a good sign that while modern science has begun to seriously investigate the mind, we do not yet have anything like a mature science of the mind and brain. Nonbelievers, in particular, cannot claim that science compels a naturalist account of the mind with the same confidence that they can point to the naturalism of modern physics and biology.

Still, it is worth noting that dualism has become a minority position even among philosophers. Moreover, though neuroscience has only very recently begun to figure out how the brain works, we know a good deal that suggests the mind is a physical process. Science, after all, is not only about the established knowledge in textbooks. Scientists are just as interested in those theories that show the best promise for unlocking mysteries. And today, materialist approaches to the mind present the best prospects for progress. Most scientists and science-oriented philosophers are convinced that their situation is similar to a treasure hunter who does not have a detailed map to the loot, but is pretty sure on which island among many the treasure is buried.

MACHINES WHO THINK

If a mind is a physical process, built out of chance and necessity, it should be possible to construct intelligent machines. An engineer

should be able to produce a robot who thinks, is aware of its surroundings, and interacts with the world in an intelligent manner. This
robot would be no mere automaton, blindly following instructions the
engineer built in; it would be an independent agent who can come up
with genuinely new ideas.

Such dreams started looking more realistic in the computer age.
Computers are information-processing devices, taking-in input, doing
computations that depend on the internal state of the computer as
well as the input, and finally producing some output. In a very general sense, this is also what humans do. We take in sensory information about the world, alter our mental state in response, and act on
the world as a result. It became tempting to think human minds were
similar to computers.

A brain looks nothing like a computer. Still, many philosophers of
mind and researchers on artificial intelligence (AI) hope that a computer analogy can illuminate some broad features of the mind. For example, in working with computers, there is a clear distinction between
hardware and software. The software is a detailed description of what
the computer *does*; most important, the same software can be implemented on different machines with different underlying hardware.
Functionalists identify the mind with software rather than hardware.
Intelligence, in a functionalist view, is in what the brain does, and the
brain's functions can also be realized on a different physical substrate.
To build a smart robot, an engineer need not try to rig up an electronic
version of a brain; other hardware will work just as well as long as
the robot has the proper software giving it the functional capabilities
defining intelligent behavior. We need not crudely identify mental
states with specific brain states.

An important reason to suppose computers might be able to think
if they have the right software is that computers are truly general-
purpose information processors. According to the famous Church-
Turing thesis, almost any process of mechanical computation can be
implemented on a computer (Boolos, Burgess, and Jeffrey 2002). If
human information processing is physical, it seems our thoughts
should have a computer equivalent in the form of the operations of a
machine with extremely complex software.

There is a mathematical wrinkle here: computers cannot do absolutely everything. There are nonalgorithmic functions no computer
can calculate. The physical world is not a place organized in neat rows
of zeros and ones like a computer memory; there is reason to suspect
that whatever our brains do, it is nonalgorithmic. However, this is not

an issue. It turns out every function can be expressed as a combination of a set of rules and sheer randomness. So a machine combining chance and necessity can perform a nonalgorithmic function, provided the task to be accomplished does not depend on any particular random sequence being produced. Randomness serves to generate novelty, producing outcomes not predetermined by any set of rules (Edis 1998a).

All this means that unless human intelligence involves some extremely magical powers (which not even any dualist claims), it can also be implemented by a machine that combines chance and necessity. Machine intelligence is almost certainly possible. Furthermore, randomness is the ingredient that gives a machine flexibility, letting it move beyond its initial programming (see Figure 4.1). A truly intelligent, creative machine will depend on processes like Darwinian variation and selection. Today, proponents of intelligent design creationism continue to claim that ID is a separate principle of explanation that cannot be expressed in terms of chance and necessity. They explicitly defend varieties of dualism about the mind. They cannot be right. It looks as if intelligent design by humans can itself can be subsumed within Darwinian, naturalistic explanations (Edis 2004b).

So there is reason to believe machines could think. Nevertheless, the work in theoretical computer science that supports this conclusion is very abstract—it says nothing about how to build an artificial intelligence. It does not mean digital computers are a good model to start with if we were to try and construct a mind. In fact, thinking in terms of a program on a familiar computer is almost certainly the wrong way to go.

Consider the famous Chinese Room argument (Searle 1980), intended to show that there was more to minds than executing a series of instructions the way a digital computer does. Imagine that say we put a man who does not speak Chinese into a room and let his only connection with the outside world be occasional Chinese characters flashed through a window. He happens to have a very extensive set of instructions for how to respond to each sequence of characters he sees; after consulting his instructions and making the appropriate computations, he selects a set of boards with Chinese characters lying around the room and flashes these at the outside world. An outside observer, the argument goes, would perhaps witness a fluid Chinese conversation going on, as in an e-mail exchange. But no one in the room actually understands Chinese—all that is happening is a series of instructions being followed.

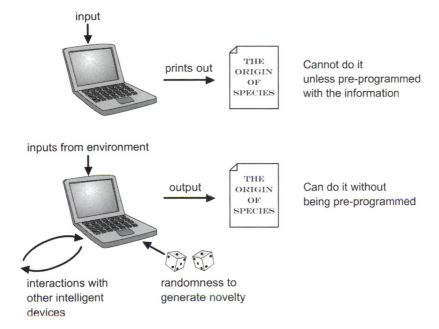

Figure 4.1 A normal computer's output is determined by its programming and input. If a machine used randomness as a source of novelty, however, it could conceivably achieve genuine, independent creative intelligence. (Ricochet Productions)

A functionalist would reply that the Chinese Room as a whole does, in fact, understand Chinese. The simple scenario misleads our intuitions; we should really imagine a very complex set of instructions that is not just written down in manuals but can grow and change the way an intelligent person learns new things. We should be able to introduce variations into the instruction set, select what works according to experience, and so on. This reply is not as plausible as it can be, however, unless the way the instructions work is fleshed out. To do this, we need to look at real language devices—human brains. And then we find that brains are not serial machines and are not programmed with any sort of code. Also, brains vary widely in structure—unlike computers, which are useful precisely because of their relatively bland uniformity of hardware. The desktop computer, while perhaps in principle capable of thought in a very broad and abstract sense, is a bad model when trying to understand actual human thinking.

For these reasons, artificial intelligence research today does not try to directly program intelligence into computers. Researchers work, for

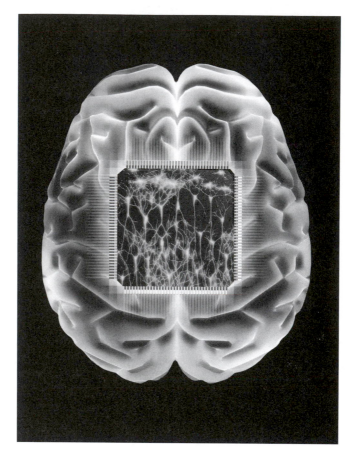

Figure 4.2   Computer artwork of a microprocessor chip contain-
ing neurons attached to a human brain. This could represent the
merging of biological and electronic networks into a form of artificial intelligence. Increasingly, artificial intelligence research is being inspired by knowledge about actual brains. (© Alfred Pasieka/
Science Photo Library)

example, with artificial neural networks,[1] constructing systems that
behave much like the circuits involved in perception and associative
memory in brains. Neural networks are not programmed in a conventional sense; they have to be trained to recognize things. AI researchers are also realizing that Darwinian processes have a critical
role to play in giving machines genuine creativity. Furthermore, AI
today is taking on broader tasks like navigating a complex world in

real time, rather than devising software that is good in very narrow task domains such as chess-playing programs. Stages on the way to machine intelligence have begun to look more like nervous systems (see Figure 4.2). If engineers eventually make smart robots, they will very likely have to be grown, not manufactured; taught, not programmed.

GENES AND MEMES

Brains are messy—the most complex objects we know of. So scientists often try and approach minds indirectly, before trying to figure out in detail what all that mass of interconnected neurons are up to.

Moving beyond the computer analogy, we first notice that human minds are not general-purpose devices. We are capable of very abstract, general-purpose information processing, formal reasoning, probabilistic inferences, and similar feats that are relatively easy to instruct computers to do. And in modern societies, these are important skills. Nevertheless, such thinking does not come easily and requires much training. In contrast, human brains are very good at some tasks that still stump computers. For example, we recognize faces effortlessly. This happens automatically, even when we only see part of a face, or when lighting conditions or viewing angles vary widely. Directly programming a computer to recognize faces—as opposed to training a neural network—is very difficult. In our brains, an immense amount of processing goes on behind the scenes to make an "Oh, there's Aunt Martha" experience possible.

Much of what goes on in our heads is not conscious, and certainly is not the operation of an all-purpose learning device. Recognizing faces is a built-in human capability; it has to be nourished in infancy, but we are born predisposed to identify faces. And it appears we have a long list of such built-in capabilities or "mental modules" (Stanovich 2004, chapter 2). Not only do we recognize faces, we also read emotional states from faces and bodies. We automatically acquire grammar as a small child. Without having to ponder over it, we classify observed objects into categories such as physical object, living thing, person, or artifact, and immediately make the appropriate inferences in folk physics or folk psychology or whatever applies to that domain. Such cognitive modules operate very quickly, efficiently, and in parallel. They directly evoke emotions and actions. They are also largely independent of one another and are specialized to particular tasks.

Though we handle them as if they were trivial, they involve very difficult computations that give engineers who would like to build an independent robot no end of headaches.

The way human brains automatically categorize sensory information is not entirely learned, not cultural, and not obvious—why should every normal human be so good at recognizing human faces but not fingerprints? Our perceptions are structured to highlight what is important to humans as a species. Our particular evolutionary history is the reason we recognize faces. And so a new discipline, evolutionary psychology, has taken an important role in understanding why human minds work the way they do.

Evolutionary psychologists have particularly emphasized the view that the human mind is built on a substrate of automatic mental modules, all of which have adaptive functions in the environment in which humans first arose. They liken our minds to a Swiss Army knife, composed of a host of independent, more or less specialized modules (Cosmides and Tooby 1994). Asking what are good reproductive strategies, they have illuminated some complex aspects of our minds, such as patterns of sexual behavior common in almost all cultures.

It may seem that such ideas have little to do with the soul, if all evolutionary psychologists are doing is exploring the "lower" or animal aspects of our selves, which religious traditions have never denied. Nevertheless, even this level of investigation raises questions about the soul. Consider morality, often thought to be an especially divine aspect of humanity. Cognitive scientists have explored how the feeling of right and wrong is tied to a kind of moral perception roughly similar to the way brains recognize faces. Among mental modules proposed by psychologists, there are those devoted to social exchange, detection of cheating, friendship, and more, which are closely related to how we take a moral point of view. Evolutionary psychologists have managed to make sense of our basic moral equipment by seeing it as adaptations critical to the survival of a social animal. These basic mechanisms do not exhaust our moral life, but it is clear that moral thought is built on a substrate of physical information processing put together by Darwinian variation and selection. What would a soul contribute to moral behavior—what is it supposed to do?

This is not to say evolutionary psychology gives us the whole picture. Our serial, logical, language-based mode of thought is not fully in command, and it appears to be a makeshift imposition on the parallel operating structure of the brain. Still, it exists and it sometimes can override more automated gut-level responses. This capability is not so directly responsive to genetic interests as are our narrowly specialized mental modules, and so it allows us to pursue other, perhaps more rational interests (Stanovich 2004). Also, evolutionary psychology has little to say about culture, that most mind-dependent aspect of our world. We do not just replicate our genes but also pass on ideas, beliefs, and knowledge.

So the next step is to think about cultural evolution and how culture shapes minds. The most ambitious approach today introduces the concept of a meme, a unit of cultural transmission roughly analogous to a gene. A meme is supposed to be an independent replicator, undergoing Darwinian evolution in its own right.

Memetics is very new, and so there is a lot open to question concerning even very basic matters such as what exactly a meme is supposed to be. For example, an advertising jingle that gets stuck in our minds and spreads rapidly may be a meme. Or the idea of the wheel. Or, for that matter, an ideology like free-market economics, though this would be better described as a memeplex, a complex of associated memes that replicate together. Some theorists think of a meme as a behavior that is copied by imitation (Blackmore 1999). Others make it a nebulous packet of "information" that floats around independent of any physical realization. Such lack of specificity leaves memetics vulnerable to severe criticism. Moreover, whatever a meme is, it is also not clear whether cultural transmission of information is accurate enough for variation and selection to take hold. If memes can only be copied with very low fidelity, they cannot properly evolve. Memeticists have been trying to address such problems; the most promising current thinking identifies memes with brain states that can, through the behavior they lead to, trigger replications of functionally very similar states in other brains (Aunger 2002). Still, memetics has to yet prove itself—in its popular forms, memetics very often makes overly grandiose claims, raising the suspicion that it is a crank notion rather than genuine science.

Even so, memetics holds some promise since humans do transmit culture as well as replicate genes. Proponents of memetics ask us to take a "meme's-eye view" and look at cultural change by asking what reproductive interests are being served. Often, the interests of the cultural replicator and its human hosts go hand in hand. The notion of a wheel or good health advice spreads largely because it benefits people. There are, however, cases where the interest served is only the reproductive interest of the meme itself. For example, a chain letter might threaten all sorts of bad luck and calamities unless copied and forwarded to six other people. And many people actually devote resources to copy it—a complete waste from a human point of view, but a successful reproduction of the meme. Since it is replication that counts, this leads memeticists to ask how minds are sculpted by memes. If brains are the living space memes compete for, then much of the activity of brains may be devoted to the dynamics of memes.

Evolutionary psychology highlights how genes constructed our brains to aid them replicate. Memetics attempts to add another replicator, consisting of certain brain states. Either way, these are materialist approaches that anchor minds in the physical world, increasingly making souls redundant.

THE CONSCIOUS BRAIN

If much of what goes on in our heads is due to automatically activated modules or replicating memes, who is in charge? After all, we think minds have a self, an "I" that perceives Aunt Martha after the face recognition circuitry does its job. And though there may be a host of memes in our brains struggling for replication, we also appear to have a coherent self, the same "I" that thinks our thoughts and has a continuous life history up to now.

It is easy to imagine this self as a kind of little person in the head. All the face recognitions, moral perceptions, and so on come forth to be presented to the seat of consciousness. The mass of independent brain processes working away in parallel come to the essential self for executive review. Dualists may be tempted to think this way, identifying the executive self with the soul. Maybe bringing it all together is just what a soul does, even if this is a somewhat diminished role compared to the more robust soul of traditional dualism.

Nondualists, however, do not have this option. A little person inside the head does not properly explain anything about a mind—it just declares it a mystery. A real explanation of selves needs to show how a continuous narrative self is cobbled together out of brain processes that, when taken alone, are less than conscious and less than fully intelligent (Dennett 1991). A rough analogy is how we can speak of "the will of the people" in a democracy without any person being a "central willer." Political decisions emerge out of the complex interactions of interest groups and constituencies without consulting the sovereign will of a king.

Accounting for the full-blown autobiographical self of a language-using animal will be very difficult. At present, researchers have many interesting ideas for how to proceed, but most of them are in the speculative, hand-waving stage. So to make progress, we need to start at a simpler level. One aspect of our minds that makes the person-in-the-head image compelling is our conscious awareness—all of what the brain does appears to come together seamlessly in a rich subjective experience of the present moment. This experience, of seeing the colors of flowers in a garden while feeling the wind and hearing birdsong, with all the accompanying emotions, seems entirely left out of the discussions of brain processes. How does modern neuroscience even begin to explain this basic, primary consciousness? After all, understanding primary consciousness may be most relevant for questions about the soul. More than intelligence or a full-blown sense of self, what seems most amazing, even magical, is the way we have a unified mental scene of the present moment. How can brain states produce conscious sensations?

Neuroscientists have learned a lot about certain subsystems of the brain, such as those areas associated with basic visual processing. They can identify circuits responsible for tasks such as detecting the edges of objects in view, and even come up with a good picture of how something like face recognition works. After this point, our knowledge gets murkier. Somehow the results of processes such as face recognition must become part of an overarching awareness of encountering Aunt Martha. We need some kind of brain state that puts results broadcast from various subsystems together in a unified fashion. Furthermore, this state should discriminate among the vast possibilities the world presents to us at different moments, noticing that not only is Aunt Martha present, but also a flower garden, a bird flying by, and a lot more. This state must be distributed throughout the brain, without invoking a mysterious little person in the head as an executive controller. On top of all this, we also have to understand how it is that we are conscious of only certain brain activities, such as face recognition, while much also goes on below the surface. We are not, for example, typically aware of what neural circuits involved in regulating blood pressure are doing.

No one yet has a complete account of how primary consciousness comes about. However, Gerald Edelman and Giulio Tononi's theory

(2000) illustrates the kind of approach that looks promising (see also Damasio 1999).

Edelman and Tononi emphasize how the brain is shaped by Darwinian-style variation-and-selection processes. First of all, during the brain's early development, many neurons die, and connections between neurons are established and eliminated. After this developmental selection sets the basic architecture of the brain, selection due to experience leads to the strengthening or weakening of connections. Two-way connections between subsystems in the brain serve to coordinate various brain activities. The brain is an extremely complex neural network and because it is shaped by growth and selection, it is more than a loose collection of specialized modules. Many brain states are functionally very similar, so no one spot in the brain or one pattern of activity uniquely stands for any kind of mental activity.

Building on the selection-shaped, flexible nature of the brain, Edelman and Tononi propose that consciousness is due to an ever-changing "dynamic core" process that enlists much of the brain. Some brain activities, such as monitoring blood pressure, are disconnected from the dynamic core and so cannot enter awareness. However, most of the results of sensory processing come together in a network that can represent an immense amount of information at any given moment. The dynamic core can therefore immediately discriminate between a vast array of possibilities presented by the world. Furthermore, the brain also links these discriminations to appropriate emotional states and actions as shaped by evolutionary history and individual learning.

The dynamic core is unified and highly integrated. The neurons involved in the core interact strongly; they cannot be separated out into distinct groups doing unrelated tasks. The dynamic core is also private; it cannot be shared by other observers. Edelman and Tononi describe how the dynamic core's properties account for many other features of consciousness: its informative nature, ability to highlight unexpected associations, the fact that we can only attend to a limited number of tasks at once, the serial nature of consciousness, and the continuous but constantly changing stream of awareness we experience. And naturally, no little person appears inside the brain. The dynamic core assembles a unified stream of experience, but it is a process distributed all over the brain.

If something like the dynamic core hypothesis succeeds—and its basic approach looks promising—it would describe how brains produce primary consciousness—the awareness of a "remembered present" that is available to a housecat as much as to humans. Brain scientists also speculate on how the brain generates a full-blown autobiographical self, but at present, ideas about the full-blown self remain less developed. Nevertheless, it seems clear to many scientists today that consciousness is becoming explainable within the physical world. Only a few decades ago, our sciences had little to say about consciousness. Today, how the brain produces conscious experience is a question that inspires research programs which are making real progress. The soul seems ever more redundant. After all, what would it do? What would a soul add, on top of what the brain already accomplishes on its own?

QUALIA AND SUBJECTIVITY

Maybe there *is* something missing from all the scientific descriptions of brain processes: felt experience. A brain process might underlie our experience of red, but there is no redness in it. Nothing in a scientific account, in fact, tells us how something feels subjectively. Many philosophers think that felt qualities of experience—qualia, in philosophical jargon—cannot be captured by any materialist description, and that qualia are what consciousness is all about.

Philosophers tell a number of stories to illustrate the point about qualia. One concerns "Mary the color scientist." Imagine a scientist who has complete knowledge about how light interacts with eyes, how the nervous system conveys visual information to the brain, and how brain states change with color awareness. Say, however, that Mary has always lived in an environment where everything is in shades of grey. Now, if Mary some day steps outside and sees colors for the first time, the experience will be new to her. She will, it seems, learn something beyond her already complete knowledge about physics and brains. Perhaps there is something nonphysical about qualia.

Another famous philosophical illustration asks what it is like to be a bat. We might some day know all about a bat's brain, map out its dynamic core or whatever brain state it is that allows a bat's conscious discriminations, and watch how this state changes. But this knowledge about the bat's brain, however complete, would not tell us what being a bat would feel like from a first-person, subjective point of view. The subjectivity of consciousness, it seems, is something no third-person, objective scientific account can capture.

Some philosophers respond to the qualia and subjectivity problems by giving up. Consciousness might, for example, be a problem that we are simply incapable of solving. Or, qualia may be attached to brain states by a strange kind of natural law over and above physical laws. Some philosophers suggest that neuroscience has not helped in advancing our understanding of consciousness, and that we should admit that folk psychology is the appropriate level of description, without seeking a deeper connection to physics. Others treat information as a mysterious entity detached from physical patterns, and look to tie it to consciousness (see Lyons 2001).

Most philosophers who are pessimistic about a scientific approach to consciousness do not dispute that mind somehow arises from matter. Or at least, they consider consciousness a natural phenomenon, even if it cannot be completely understood in physical terms. So problems about qualia need not challenge a naturalistic view of the world, and hence do not bother too many nonbelievers. Still, setting qualia aside as an exception puts a strain on a bottom-up description of the world.

One immediate problem with nonphysical qualia is how they are supposed to affect physical objects. If Mary steps outside and sees color, we expect that her experience of color changes her aesthetic appreciation and her behavior when beholding the now color-filled world. If subjective qualia are the cause of all this, then they look very much like they are made of ghostly soul-stuff. The problems Cartesian dualists had with souls interacting with the physical world surface again. So philosophers typically assume qualia have no physical effects: everything that happens in the physical world is fully accounted for by the physics and brain science of which Mary already had full knowledge. But in this case, qualia only accompany experience, without participating in any of the behavior we see in aesthetic appreciation or anything. This is a peculiar conclusion. Many philosophers worry that the scientific image of persons somehow diminishes us, and they seek hope for human dignity and significance in mysterious qualia unexplainable by science. Causally impotent qualia are not very likely to satisfy this hope.

A more attractive solution could be to resurrect old-fashioned dualism, and some religious philosophers take this path (Habermas and Moreland 1998). They argue that qualia and subjectivity mark the breakdown of a materialist approach and then go further by attributing causal powers to the soul. We are supposed to have godlike free will, being an uncaused cause of our decisions. They say certain experiences, such as those related by people who, when near death, perceive their souls separating from their bodies and traveling down a tunnel of light, are firsthand experiences of the mind existing apart from the body. Other people demonstrate psychic powers, showing that mind goes beyond matter. And recognizing the reality of the soul might solve problems facing scientists, such as assembling a unified stream of experience from all the different processes going on in the brain.

The dualist option, though, is similar to creationism in biology. Indeed, defending dualism and opposing materialist views of mind is one of the major preoccupations of the intelligent design movement. Dualists exaggerate the difficulties mainstream science faces, portraying problems as insurmountable hurdles while at the same time ignoring the real progress made in solving them. They rely on dubious evidence drawn from the disreputable fringes of science—in this case, parapsychology. They promise mystery rather than explanation.

Consider Mary the color scientist again, now through materialist eyes. When she encounters color for the first time, her brain will take on a state it had never been in before. All her previous knowledge about brains and light had not put her brain into a state of experiencing color. She does not, in fact, acquire any new information about the world. There is a substantial physical difference, however, between her brain states before and after she sees color. An image file on a computer disk contains the same information whether someone looks at the 0's and 1's that make up the file or whether they properly display the encoded image on a screen (see Figure 4.3). Encountering color similarly re-presents information Mary already had. The difference is profound, but it is also a physical difference in her brain.

We can, for that matter, know something about how it feels to be a bat. After all, the bat's dynamic core, or whatever the states underlying the bat's primary consciousness may be, have some similarities to the states available to human brains. Now, there is no way to exactly reproduce the conscious state of a bat in our own brains because our species also have profound differences. So there is no way for us to experience firsthand exactly what being a bat feels like subjectively. But this impossibility is due to those sorts of brain states not being entirely available to us. There is no mystery here: the privacy of con-

On Screen   **In Memory**

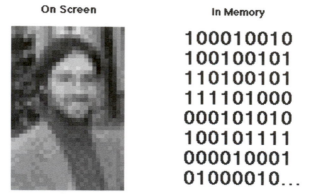

100010010
100100101
110100101
111101000
000101010
100101111
000010001
01000010...

Figure 4.3   The image on the screen and the raw bit pattern of the image file contain the same information. But they do not lead to the same response in an observer. How information is represented is *important*.

scious states and their inability to be reproduced exactly in all observers is easily understood in a materialist framework.

We cannot expect even a fully accurate and detailed scientific description of a conscious brain to reproduce a particular brain state and the subjective experience with which it is associated. No scientific description generates the experience of what it describes. A description of a physics experiment can never substitute for performing the experiment. The difference is that physicists can use the third-person descriptions of physics to set up a public experiment that generates similar conscious states in observers. For a neuroscientist, reproducing a first-person state would have to involve direct intervention in brains.

So to a materialist, problems with qualia and subjectivity dissolve into a matter of switching between first-person and third-person points of view. Qualia are, literally, the discriminations available to the conscious states of brains. If we can identify a dynamic core or a similar state in a brain, we know it has to have a first-person point of view associated with it. From that point of view, which outsiders can never fully adopt, qualia are colors, sounds, and other sensations. Nevertheless, these first-person qualia are not something else on top of the discriminations described from a third-person point of view. The brain states producing the sensations interact with the rest of the physical world of which they are a part. Indeed, the evolutionary function of consciousness is to provide animals with rapid discriminations the organism can act on. It is not something mysterious that does not participate in the world (Edelman 2004).

MIND OVER MATTER

The kind of dualism associated with mainstream religions is not the only alternative to materialism. Another option is to take consciousness itself to be the fundamental reality. Indeed, the occult spiritual tradition conceives of the universe as a living being permeated by intelligence. And today, New Age thinkers claim that their meditations on mystical matters provide the best insights into the nature of consciousness, and that modern science, particularly physics, confirms an occult picture of the universe.

Alternative spiritualities often incorporate a strong moral criticism of religious orthodoxy. They portray organized religions as authoritarian institutions devoted to ossified dogmas, more likely to stifle spiritual impulses than to give a seeker direction and discipline. Such a moral stance is attractive to many nonbelievers. Moreover, New Age ideas about the nature of reality reinforce moral objections to orthodoxy. The God of individualist spiritualities is often a nonauthoritarian, only fuzzily understood spiritual force; it easily becomes an almost impersonal universal consciousness. The natural and supernatural gradually shade into each other; spiritual and material realms are not sharply separated. In the same vein, New Age thinkers condemn mind-body dualism. The traditional Christian exaltation of soul over body is a favorite New Age target, blamed for many evils including undervaluing women, sexuality, and the natural environment.

Historically, occult attitudes toward science have been ambivalent. Many New Agers are superficially very positive about science and take care to argue that they present a synthesis of scientific and spiritual views that can heal the rift between science and religion. Even so, to New Agers, the skepticism modern science engenders about spiritual realities comes across as a kind of dogmatic blindness. Also, thinkers sympathetic to alternative spiritualities often describe mainstream science as being committed to mind-matter dualism (Burton and Grandy 2004). If so, this would mean that science suffers from the perennial dualist problem of understanding how spirit and matter interact—scientists ignore the question of how it is possible to gain scientific knowledge at all. If consciousness is fundamental, however, we gain knowledge by participating in a living universe. Furthermore, we can be open to other ways of knowing, such as the spiritual wisdom obtained through mystical and paranormal experiences. Consciousness can make contact with reality directly—creating and influencing matter by an act of will—as well as through the medium of laboratory experiments.

So the occult stream of thought, though attractive to nonbelievers who morally object to orthodox religion, is really a third option, a rival to both mainstream science and conventional religion. Science-inspired nonbelievers tend to lump alternative spirituality in with other religions; after all, the New Age subculture also adopts a top-down, spiritual view of the world. And in the context of the debate over minds, New Agers, though they object to the sharp, Cartesian version of dualism, still defend a view much closer to the traditional humanist image of persons. Nevertheless, the occult view of mind is worth considering in its own right.

The occult option so far has had little influence on the scientific effort to explain minds. One reason is that while occult thinking can produce inspiring spiritual visions, it has trouble making contact with mundane, testable realities. The closest thing to an antimaterialist research program concerning minds has been psychical research, now called parapsychology, which since the nineteenth century has been on a quest to demonstrate superpowers of the mind and discover that souls can survive death. Parapsychology, however, has never been able to produce the sort of results that could convince a skeptical mainstream scientific community of the reality of paranormal phenomena. The consistent story with parapsychology is a complete lack of theoretical development, and the lack of any data that stands out from among the sort of mistakes and hoaxes that can be expected from a messy, complicated experimental environment involving human subjects. Parapsychology today inhabits a gray area between legitimate science and magical thinking, and few scientists investigating the mind expect it to ever contribute anything of substance.

Defenders of a broadly occult view of the mind often claim support from modern physics, alleging that a form of consciousness that is not explainable in terms of other physical interactions is integral to quantum mechanics. Some suggest consciousness is an all-pervading field analogous to the quantum fields that inhabit the fundamental equations of physics. In fact, the early development of quantum physics included some speculation about the role of conscious observation.

Ironically, the reason in part was the Cartesian dualism about the mind that was still characteristic of early twentieth-century science.

It is possible to find physicists, particularly physicists who adhere to Eastern religions, defending such views (see, for example, Goswami 2001). Still, while such efforts might enrich the stock of metaphors used by religious believers, they have been singularly unconvincing to the scientific community at large. Cognitive and neuroscientists are dubious that the physics of elementary particles and forces has anything to contribute beyond setting the stage for probing the complexities of the brain. And most physicists are skeptical that consciousness has any fundamental role in quantum physics, other than observation being a complex macroscopic interaction with quantum states. "Quantum mysticism" is good for sales in the metaphysical sections of bookstores, but it has a bad reputation in the scientific mainstream.

The idea of tying consciousness to quantum physics occasionally emerges as a legitimate proposal. For example, Roger Penrose (1994) argues that true intelligence cannot be algorithmic, that quantum phenomena associated with certain neuronal structures may be responsible for the unity of consciousness, and that a proper theory of quantum gravity might be needed to understand both consciousness and quantum measurement. Some of Penrose's arguments have merit, as with his observation about nonalgorithmicity. These can easily be accounted for, however, within the mainstream of machine intelligence research (Edis 1998a). Penrose's other speculations have had no significant effect; they are generally considered unlikely at best.

Though mainstream cognitive and brain sciences have just begun to make substantial progress, and we have immense gaps in our understanding, they have no real rivals, nothing that has the same prospects of improving our knowledge of the mind. The occult third option seems as much a dead end as more orthodox forms of dualism.

THE GOD MODULE

Another popular way to reconcile science and religion is to argue that current science actually supports ancient beliefs. As with modern physics and evolution, more liberal religious thinkers contest the use of neuroscience to support nonbelief.

At face value, denying the soul is not religiously promising. If all experiences come about through brains and their interactions within a physical world, the supernatural no longer seems to have much to do. But this leaves open the possibility that religious experience accurately testifies to some sort of transcendent reality. Maybe our material brains have a built-in capacity to experience a spiritual realm.

Modern defenses of religious belief rely heavily on religious experience. Spiritual experience may include the quiet convictions of an ordinary churchgoer who perceives God at work in her life, but arguments from religious experience usually emphasize the spectacular reports from mystics and visionaries. These experiences give us, devout philosophers are apt to say, a direct acquaintance with the divine. There is no more reason to doubt such spiritual perceptions than to doubt the veracity of our direct acquaintance with everyday reality. The divine reveals itself to those who are receptive.

Nonbelievers are less impressed with religious experience. The diversity of doctrines affirmed through mystical experience, for example, means that religious claims to direct knowledge of ultimate realities are suspect. When a Christian mystic directly experiences the Trinity and a Buddhist mystic comes to know that emptiness underlies all, we have to think religious experience is very much conditioned by the culture of the believer (Katz 1983).

Today, nonbelievers also look to psychology and brain science to add to their longstanding philosophical doubts about religious experience. As the brain processes involved in spiritual perceptions are better understood, it becomes less tempting to seek a divine source for even the most sublime experiences. For example, the following observations demystify mysticism:

- Mystical states come in two broad categories: a more introspective sort that invites believers to encounter the divine in a deep quietness, and an extroverted, visionary mysticism. Both usually involve minimizing input from the external world: introspective mysticism takes place in a state of very low stimulation, while visionary experience happens when the brain becomes highly stimulated in isolation from the external world.

- Religious traditions that value mysticism develop techniques to induce mystical awareness. Whether it be through extreme bodily circumstances, continually recycling a mental subroutine, or otherwise trying to detach awareness from outside objects, such techniques drive the mystics' brains into extreme, unusual states.

- Certain drugs, diseases, or exposure to strong magnetic fields, and other interventions in appropriate brain areas can also produce experiences

within the standard mystical repertoire. These include a sense of timeless-
ness or time distortion, the fading of the perceived boundary between the
self and the outside world, encountering an overwhelming personal presence, and more.

- Some areas of the brain are especially important for emotions, time perception, a self-boundary, person recognition, and so forth. Researchers can,
in laboratory settings, disrupt their normal operations and produce mystical experiences. They can even study the brain processes involved
through brain-imaging technologies (Edis 2002, chapter 7).

Progress in understanding religious experience, however, does not
automatically support nonbelief. Religious thought is very adaptable,
and "neurotheologians" have leapt into the breach to propose supernatural interpretations of brain science. Mystical experience is produced by a material brain, like all experience. But just as our brains
are well-equipped to negotiate the everyday world through normal
experience, perhaps the mechanisms underlying religious experience
means that human brains are equipped to make contact with an "Absolute Unitary Being" (Newberg, d'Aquili, and Rause 2002). Perceptions of and belief in supernatural realities is near-universal in
humans, so it might make sense to speak of human brains possessing
a "God module." In that case, like other modules of the brain, the God
module should have an adaptive function. And the most natural function would be to respond to an actual spiritual reality. In other words,
God made our brains such that we are able to perceive God.

As it turns out, the detailed arguments proposing a God module or
a built-in capacity to recognize the divine run afoul of much recent
cognitive and neuroscientific research (Atran 2002, chapter 7). It is true
that humans are predisposed to perceive the world in supernatural
terms, but this does not require specialized structures in the brain devoted to the gods.

Mystical experience in particular does not have such a significant
role in ordinary religious belief. Mystic raptures come about in rare
circumstances, and are typically available only to a few religious specialists. And their experiences are shaped by their already existing religious traditions—they cannot be taken as direct contact with reality
unadorned by interpretation. Furthermore, there are traditions such
as some forms of Buddhism that incorporate a more sophisticated introspective psychology than those developed within Western religions. Buddhist meditators testify to going *beyond* experiences such as

unity with an overwhelming person, finding that the self breaks down as all is revealed to be impermanent, in flux. The nontheistic psychology of Buddhism is much more in line with modern cognitive and brain science than the traditions of the theistic religions.

The rarity of the unusual circumstances producing mystical experience also makes it unlikely that the capacity for mysticism has any notable selective value in reproduction. Any advantage this capacity confers must be indirect—a person who has experiences easily interpreted as contact with the divine will be considered a respected holy person in many cultures. But if this is so, the religious culture and a general propensity to perceive supernatural agents as responsible for the world must already be in place. Mysticism fits religion well, but it is not a major cause of religiosity. There is no "God module" in the brain analogous to modules serving perceptions of external reality.

In any case, religious experience is not very similar to ordinary perception. Mystics are typically passive during their experience, and while they produce verbose reports of their visions, they convey virtually no reliable information. The basic problem with neurotheology is that brain science has made progress explaining mystic episodes without bringing in anything beyond the brain. Unlike normal perception, no external, independent source of information seems to be needed to account for the details of mystical experiences.

PARTS AND WHOLES

The tendency in science today is to see the mind as a physical process, without any mysterious souls involved. Researchers take a reductionist approach, expecting that everything about life and the mind ultimately is realized by physical particles and interactions that are lifeless and mindless on their own (Melnyk 2003; Weinberg 1992). Complexity emerges from the bottom up; it is not imposed from the top down by a nonphysical spiritual reality.

The bottom-up perspective of modern science does not mean that everything interesting takes place among subatomic particles, or that physics is the only worthwhile science and the others are just ways of working out the consequences of the basic laws of physics. Still, defenders of the humanistic image of persons often worry that without souls we become nothing but a pointless collection of particles. They think that to preserve human dignity, we must have some kind of top-down causation or some way in which the whole is greater than the sum of its parts.

Seeking magic in "wholes," however, overlooks how most of the interesting research in science has little to do with fundamental physics. Physics sets the stage, but cognitive and brain science today look to explain the mind in terms of extremely complex but ordinary physical interactions—not new and exotic fundamental physics. Even most of the work done in physics starts not with subatomic interactions but by identifying the appropriate higher-level objects. An atom, for example, is a coherent entity with its own integrity, and a physicist studying how a gas of atoms becomes liquid at a certain temperature does not need to know everything about how subatomic particles come together to make an atom. Now, a fuller understanding of the atom does require explaining how atoms form, and physicists solve this problem in a bottom-up fashion, with no mysterious wholeness involved. The higher-level description of a gas in terms of atoms is anchored on the lower-level descriptions of electrons, nucleons, and photons. Nevertheless, physicists studying the gas-liquid transition do not translate their explanation into an account of what the electrons, nucleons, and photons are doing as the liquid forms. Such an account, though possible in principle, would be useless. In explaining the gas-liquid transition, the appropriate description treats atoms as entities in their own right—a lower-level description would drown the interesting features of the physics in a mass of irrelevant detail.

A higher-level description is best because it can summarize large amounts of information about strongly interacting components. And so it is very important in science to identify natural kinds—to carve up the world in terms of objects with real integrity, that we can manageably understand. For example, living things can be understood in terms of their chemistry. However, they are not arbitrary collections of organic chemicals. And an ecologist, working at the level of organisms and species, does best by ignoring most of the chemical details.

Because science often approaches complex systems by breaking them apart and looking at the interactions of the parts, scientists are often accused of ignoring wholes, oversimplifying the world, or jumping to the conclusion that the world is merely physical only because they can explain some systems in a reductionist manner. None of this is true. Consider how living things affect their chemical environment—for example, because of life, the atmosphere of our planet is very different than lifeless planets. A good model of the atmosphere will include the physics of gas circulation and the absorption and emission of radiation, the complex chemistry of the atmosphere, and biological effects like those due to plant cover on the surface. All of these components of the model affect one another. Burning large swaths of forest will affect the radiation because of the change in plant cover and the large amounts of carbon added to the atmosphere, which in turn will affect how the forests will respond. Many scientists today work with extremely complex, strongly interacting systems. They cannot predict how the system will behave by a simple examination of its parts—they need to build a complex model. Identifying the correct parts and interactions, however, is critical to understanding a complex system. Scientists seek not just to reproduce how a system like the atmosphere behaves, but to explore how its behavior changes when its components are manipulated. If, say, we want to find out how industrial byproducts affect global temperatures, we have to have the correct parts and interactions in our atmospheric model.

Vague complaints about parts and wholes will not do—science simply has been too successful explaining complexity by a bottom-up approach. So some theologians (see, for example, Brown, Murphy, and Maloney 1998) profess to accept modern science completely, and even acknowledge that the mind is a physical brain process, and yet defend a version of mind-first, top-down causation.

In complex systems, a version of top-down causation may seem to take place. For example, since life shapes Earth's atmosphere, we can think of higher-level biological causes affecting lower-level chemistry. This is, however, misleading. Biological causes can be resolved into physical processes: there are no ghosts, no life forces on top of normal physics. The composition of our atmosphere is best understood by including biological-level causes because translating everything into chemistry would produce an overly detailed description that would obscure rather than illuminate. A good overall explanation of our atmosphere will be complex and multilevel; nevertheless, it remains a fundamentally bottom-up description. There is no top-down causation independent of the physics.

Similarly, though minds happen in soulless brains, it is perfectly legitimate to speak of mental causation. The brain processes that underlie beliefs and desires lead to changes in the world. But again, all of this appears today as if it can be resolved into physics. Minds have physical effects, but this is not surprising, as minds are realized by physical brains. Nothing in our knowledge vindicates the notion of a hierarchy of levels stretching up to a divine reality that manipulates the world through a form of top-down causation.

NONBELIEF ADAPTS TO BRAIN SCIENCE

As scientists increasingly think minds are material processes, science-based arguments for nonbelief become stronger. Ghosts, souls, spiritual powers—these are the very stuff of supernatural belief. Though conceptions of divinity vary widely, supernatural agents are imagined to be bodiless spirits analogous to our own minds. If human minds emerge from the world of physics, how can we even begin to think about disembodied agents? And gods are supposed not just to act on the material world, but to act intelligently. If intelligence itself is found to be a material, Darwinian process, how can we think of intelligent design as a principle separate from chance and necessity? Materialist conceptions of mind do not merely cast doubt on hopes of personal immortality or on old-fashioned gods, they challenge the very idea of a spiritual realm over and above the material world.

The notion that the mind has a material basis has not entirely sunk into our popular culture. Most significantly, neuroscience has not inspired anything like the opposition to biological evolution, although it also strongly challenges religious conceptions of humanity. Perhaps this is because the modern sciences of the mind are too new. Or maybe the dualist, humanistic image of persons is too deeply ingrained in our cultures, and it is easy to keep that cultural background undisturbed while superficially acknowledging some physical effects on our souls. That mental illness can respond to drugs, or that even more or less normal people can alter their moods with Prozac does not raise too many eyebrows. Even religiously conservative people have come to accept that much mental disorder has physical causes rather than being due to malign spirits. On the other hand, Pentecostal-style spirit-possession religions are as popular as ever, and belief in a soul surviving death remains very strong. Many intellectuals wring their hands about scientific challenges to the humanistic image, but clearly the general public is only very dimly aware of any such challenge.

To further complicate matters, the emerging scientific image bothers not just religious thinkers but also many nonbelievers. Many object to religion for moral reasons, and the desire to assert human free will over the alleged will of an authoritarian God motivates many to question churches and prophets. Popular moral polemics against religion are almost always conducted in terms of the humanistic image of persons. Controversies rage over the fairness of predestination to heaven or hell, or over the moral qualities of holy writings. The mind's dependence on the brain does not help score points in such debates. If anything, defenders of religion have the upper hand, because our everyday moral thinking depends on the humanistic image of persons. For example, we generally assume that we possess a godlike free will, on which basis we and we alone have responsibility for our uncoerced actions. Without free will and responsibility, we think, moral praise and blame do not make much sense. But if minds are natural processes, a combination of chance and necessity like all else, godlike free will cannot exist. It seems if our actions are determined by natural causes, we could not have decided to act otherwise. And the element of chance does not help: what difference would it make if subatomic throws of the dice affected the decisions made by a brain? Popular religious thinkers argue that the science-inspired, materialist kind of nonbeliever denies moral responsibility, and thereby makes the nonbelievers' own moral criticism of religion nonsensical.

Philosophical naturalists who are impressed with today's science of the mind, then, are often drawn to the debate over free will (Dennett 2003; Flanagan 2002). It seems people are thinking machines, souls and spirits do not exist, and we do not have godlike free will. But if so, what happens to the moral and cultural ways of thought that are as important intellectually and practically as natural science? What kind of free will do we in fact have, and how must our notions of moral responsibility change? These are knotty philosophical questions that are only now being properly posed, let alone answered in a way that satisfies many. Nonbelievers, who cannot resort to spiritual concepts to cut through the knots, face these problems in a more urgent manner.

NOTE

1. Neural networks are, interestingly, typically not built but simulated on digital computers. The computer is a very good general-purpose machine; though its hardware can be misleading when thinking about the mind, it is very suitable for functionally realizing alternative hardwares.

Chapter 5

⟠

The Fringes of Science

ULTIMATE EXPLANATIONS

Why is the world exactly as it is, and not otherwise?

A favorite religious answer is that in the end, it is because of the divine will. Some believe that human sin has interfered with the divine purpose behind our existence, such as those creationists who trace all death and decay, even the existence of the second law of thermodynamics, back to the freely willed disobedience of the first human couple. Nevertheless, ultimately, the cause behind everything is supposed to be God.

Metaphysical thinkers work toward the same conclusion. Everything appears to have a cause, and those causes have yet other causes. But physical causes within the world cannot account for why a different world, with a different chain of causes, could not have existed instead. The ultimate explanation for why our world is as it is must be the will of God. In fact, the classic cosmological argument for God just says that there must be an ultimate cause, and identifies this first cause with the theistic God.

Metaphysicians do not like science intruding on their turf. Science, they say, properly confines itself to unveiling "proximate causes" for what makes the sky blue or how a stomach digests food. About ultimate causes, science must be silent. Unfortunately, natural science complicates the question about an ultimate cause. The ambitious nature of modern science inevitably leads it into what used to be metaphysical territory.

Physics suggests that the answer to why the world is exactly as it is, is "why not?" This seems frivolous at first. Modern physics and cosmology, however, highlight the role of chance in the universe. Even our most basic physical laws only tell us what kind of dice were rolled to generate our universe. Everyday cause and effect is itself not fundamental, arising instead from the random interactions of microscopic physics. Physics allows for a very wide range of potential universes; it does not determine the exact history our universe should have. So a response of "why not" is very serious: it emphasizes how looking for an ultimate cause for everything is asking the wrong question.

Biology also deflates the quest for ultimates. We might, for example, ask why a cat's eyes are set close to each other, facing forward, while a rabbit's eyes are more to the side of its skull. An immediate explanation is that during the development of rabbit or cat embryos, guided by their DNA, their eyes are built differently; we might even try to roughly describe much of the intricate chain of chemical interactions involved. The ultimate explanation, however, is evolutionary adaptation (Mayr 1991). For a predator, the superior depth perception afforded by closely set eyes is advantageous; for prey, it is more important to have a wide angle view to detect danger from all directions. Natural selection will favor genes appropriate for a predator or prey way of life. A third eye in the back could be useful for both rabbits and cats, so why does it not appear? Evolution is constrained to work with what is available. Radical changes in body plans are extremely unlikely to happen for already complex animals. And that is all. An ultimate explanation in biology picks out an evolutionary history produced by chance and necessity, driven by blindly replicating genes.

Even today's sciences of the mind cast doubt on metaphysical intuitions. Traditional metaphysics is dualistic, and thus easily proposes the will of some supernatural agent as an ultimate cause. But if we take a naturalistic approach, our own minds must be very different than what dualists have imagined. It becomes hard to think of a disembodied will that is an uncaused cause—after all, our own minds are nothing of the sort.

If all this is correct, science looks more and more like it can stand alone in explaining the world. Nonbelievers inspired by science usually say that therefore nonbelief is most reasonable, that naturalism sticks to the science and merely refuses to add unnecessary supernatural entities to a very well supported scientific picture of the world.

When nonbelievers enlist science in their cause, religious thinkers often suspect that their talk of science is a disguise for a preconceived materialist philosophy. Creationists, for example, charge evolution with being an atheistic, faith-based philosophy, not a conclusion based on independent evidence. Theological conservatives are not the only ones who see prior metaphysical commitments where nonbelievers claim science. Accusing science-inspired nonbelief of illegitimate "reductionism" or "scientism"—assuming natural science is the only model for genuine knowledge—is common religious fare.

Such accusations, however, usually depend on an overly narrow conception of how science operates. Scientists do not proceed according to a preset "scientific method," but try to figure out how things work as best as they can with whatever tools they have. As physicist Percy Bridgman put it, there really is no scientific method, merely individuals "doing their damndest with their minds, no holds barred" (1947). Science is not set in stone; its methods can change and adapt to new problems. And clearly, modern science is ambitious, and so far spectacularly successful. In the face of this success, trying to carve out a separate realm for religious truth, making it immune to scientific criticism, comes across as special pleading. Isolating religion from science also ignores how many religiously important claims appear to be fact claims open to scientific investigation. Even as a practical matter, setting religious "truths" apart from science is dangerous. Compared to the tangible accomplishments of science, religious truths start looking like a second-class species of truth, if they are really true at all. God turns into a cosmic Santa Claus.

Modern science's challenge to spiritual beliefs cannot be avoided by philosophical maneuvers. What could work instead is to challenge science-based nonbelief on its own turf. After all, naturalists claim that scientific investigation will help us construct the best overall big-picture explanation of the world. If we could find some anomalies, some exceptions that break the bounds of physics-inspired naturalism, then it would become clear that the naturalistic picture is incomplete. Science as practiced today might be successful in its proper domain, but it may well need to expand its horizons to include spiritual realities if it is to encompass such anomalies. There may be more things in heaven and earth than dreamt of in naturalistic philosophy.

MIRACLES AND REJECTED KNOWLEDGE

Phenomena defying science should be easy to find, or so it seems. Popular religions from all over the world have always produced plenty of wondrous stories (McClenon 1994): the extraordinary feats of saints and gurus, scriptural prophecies, Muslim farmers slicing a vegetable open to find one of the names of God inscribed in Arabic. And then there are the miracles of traditions outside religious orthodoxy. Psychic mediums communicate with the dead or bend spoons with the power of their minds. Occultists gaze into the future. Alternative healers do everything from give dietary advice to magically remove cancer.

In fact, while most among intellectual and educational elites continue to respect modern science, spiritually inspired rebellions against science are everywhere. The advance of science has had little effect on belief in Marian apparitions among Catholics or in the magical powers of Sufi masters among Muslims. That Jesus was resurrected is very doubtful according to modern critical history (Price and Lowder 2005), but evangelical Christians continue to think that the resurrection is a historically well-attested event. The occult tradition flourishes in the form of belief in psychic powers and the amorphous New Age culture. Many think aliens visit Earth in flying saucers, abduct people, and try to convey a spiritual message to humanity.

There is more. Astrology, crop circles, iridology, the Bermuda Triangle, homeopathy, the Loch Ness monster, recovered memories of past lives—the list is endless. The variety of popular beliefs that grossly contradict the present state of scientific understanding is overwhelming. There is little in common among all these beliefs besides their being "rejected knowledge," considered dubious by intellectual and educational elites. The classic UFO, for example, is supposed to behave in ways that violate our present understanding of physics, but the favored unconventional explanation is an alien spaceship. Belief in aliens need not have anything to do with psychic powers, religious miracles, or alternative medicine. Likewise, the constituencies for different forms of rejected knowledge can be very different. Many conservative Christians, for example, favor creationism over mainstream science but have no sympathy for astrology or New Age beliefs.

Very often, however, even rejected beliefs that seem to have no religious relevance acquire spiritual overtones. Alien space travelers using advanced technology become modern-day equivalents of angels and demons (Lewis 1995). Bigfoot goes beyond being a large animal unknown to zoologists; some believe it to be an elusive magical creature exposing the limits of scientific materialism. Furthermore, just the fact of being rejected can forge connections between different fringe and paranormal beliefs. Creationists can accept psychic powers and UFOs, if they come to see them as demonic manifestations. It is not difficult to find UFO believers who say aliens display paranormal abilities.

So even though paranormal and fringe-science claims are a motley bunch, as a whole they present a similar challenge to scientific naturalism. They are miracle stories. And the reality of miracles would mean that naturalism is incomplete at best. The traditional spiritual perception of the world, as a top-down and purposeful place, could be restored (see for example, Griffin 1997). Best of all, miracles would bolster faith through tangible evidence, not by the speculative metaphysical arguments science-minded people find so easy to ignore. Science would have to expand its horizon and become an enterprise that supports spirituality instead of continually casting doubt on the gods.

The paranormal and the miraculous challenge science and nonbelief even more strongly in the cultural arena. Though mainstream science dominates formal education, in the process it also ends up appearing stodgy and dogmatic—bookbound rather than open to new experience. The socially conservative distrust science because it can threaten morally significant myths. Miracles reaffirm the true faith. But those with more antiestablishment attitudes need not adopt a scientific skepticism about the paranormal. They very often see science as yet another orthodoxy of belief, as narrow-minded as the dogmatism of organized religion. Miracles signify that the universe is open to spiritual influences, with boundless possibilities for human freedom.

Scientists, then, often worry about popular forms of rejected knowledge, as their very popularity is a sign of the cultural weakness of science. Even after successfully passing many science courses, students very often do not absorb a scientific outlook. It is not rare to find medical doctors rejecting evolution or high-technology innovators pursuing New Age ideas. At the least, most scientists look at widespread belief in paranormal and fringe science claims as a failure of science education. Many worry about the health of the scientific enterprise in a culture where healing cancer by the power of prayer seems an attractive and reasonable possibility.

Nonbelievers also have reason to worry about the popularity of un-scientific claims. Historically, science-inspired nonbelievers have hoped that as the scientific outlook spread and deepened among the public, spiritual beliefs would fade away. This is clearly not happening. Moreover, the partial success of science in modern cultures has not always helped the cause of religious doubt. Science has become part of the establishment, an institution that flourishes by providing services to powerful military and commercial interests. Morally motivated doubt about religion, however, usually has an antiestablishment flavor: rebelling against the gods has been a bid for human freedom. If science comes across as an establishment institution dogmatically rejecting paranormal claims, moral discomfort with orthodox religion can more easily lead to alternative spiritualities rather than disbelief in the supernatural.

Debate over paranormal claims does not directly involve God, revelation, or ultimate meanings. Still, this debate is very important to the relationship between science and nonbelief today. Science-inspired nonbelievers spend much of their energy promoting skepticism about paranormal matters.

SCRIPTURE AS MIRACLE

Today, a much larger percentage of the world's population is literate compared with a few centuries ago. With higher literacy, and the breakdown of peasant society and traditional forms of economic production, religions have also changed. Especially in Judaism, Christianity, and Islam, scripture has become more available to the masses. Revelation is no longer mediated exclusively by religious specialists. Scriptures written thousands of years ago can play a more direct role in popular religion today.

As established religions have become more democratic in their use of scripture, the way they perceive supernatural action in the world has also changed. Fundamentalist movements are usually suspicious of freelance magic and the divine authority claimed by gurus and saints—scripture should be sufficient. Many Protestant sects have had a rationalist tendency, restricting miracles to biblical times. Today's Islamic revivalists often denounce folk religiosity and Sufi mysticism alike as superstitions that cloud the clear guidance available from the Quran. But if scriptural fundamentalism is not to be an inadvertent force for secularizing society, it must retain a sense of magic. The divine must intervene in our lives.

Very often, the solution is to emphasize how scripture itself is a miracle. After all, whether the Bible, the Quran, or the Vedas, the local scripture is supposed to be the divine word. In another nod to modern times, an impressive way to validate the miraculous nature of scripture would be to present public, objective evidence that scripture was not a human product. If science could be enlisted to support the claims of scripture, this would be best of all.

Scriptural literalists, then, often try to show that natural science and critical history are compatible with scriptural accounts, down to the most striking miracle stories. Among the most ambitious ways to defend such a literalism are claims that the Bible or Quran, though they have been written thousands of years ago, proclaim scientific facts that are only now being discovered. If true, such claims would clearly and effectively demonstrate a superhuman hand in the scriptures.

Among Christians, young-earth creationists are among the leading proponents of claims that the Bible miraculously contains scientific knowledge. Creationist literature finds biblical mention of everything from the fundamental conservation laws of physics to dinosaurs—the "behemoth" described in Job 40:15–24 (Morris 1993). The most extravagant scripture-miracle claims, however, come from the Islamic world. Muslim apologists have an extensive literature claiming, for example, that verse 51:47 in the Quran, "We have built the heaven with might, and We it is Who make the vast extent," tells us about the expansion of the universe following the big bang. They go on to claim that verses 23:13–14 and others anticipate modern embryology, that verses 55:19–20 refer to a salinity barrier between the Mediterranean sea and the Atlantic ocean, and much more. All such claims rely on forced interpretations and dubious translations; none of these verses could have been used to predict the alleged scientific phenomena they describe. Even legends about scientists converting to Islam are common across the Muslim world. Popular but false stories include oceanographer Jacques Cousteau converting to Islam when he encountered the alleged salinity barrier verses, or that astronaut Neil Armstrong heard strange sounds on the moon that he later discovered were the Muslim call for prayer and so also converted.

Critics of such claims generally wave them off as palpably absurd, adding that tortured interpretations of scripture verses to make them

sound scientific do no good for either science or religion. More sophisticated religious scholars also tend to be bothered by such popular apologetics. First, such efforts are so embarrassing that they harm
the standing of religion among modern intellectuals. And second,
many religious thinkers are disturbed by the whole idea of validating
scripture by making it fit current science. Such an approach seems to
take the changeable claims of science to be more reliable than what
were supposed to be divine verities.

Nonbelievers, naturally, are especially severe in their criticism of
science-in-scripture apologetics. They also go further and argue that
scriptures are full of scientific mistakes. Where popular religion centers on magic and scripture, popular nonbelief will include arguments
that magic is delusion and scriptures are riddled with errors. Taken at
face value, the Bible and Quran contain plenty of internal contradictions, which are constantly highlighted by nonbelievers (see for example, McKinsey 2000) and excused away by religious apologists. The
historical narratives and alleged prophecies of the Hebrew Bible
(Callahan 1997) are also a favorite target of nonbelievers, who point
out how the Bible often does not square with our historical knowledge. Catalogs of scientific mistakes just add to nonbelievers' lists of
errors in scripture.

Finding scientific mistakes in scripture is not hard. For example, the
Quran clearly relies on ancient Near Eastern and later Hellenistic mul-
tilayered pictures when describing the universe, speaking of the seven
layers of the heavens in verses 2:29 and 41:12. Far from miraculously
anticipating modern knowledge, the Quran takes archaic cosmologies
as its background when proclaiming the power of God. The Bible gets
its biology wrong, saying rabbits "chew the cud" (Lev. 11:5–6) and en-
dorsing the ancient folk belief that seeds must die before they germi-
nate (1 Cor. 15:36). The Genesis story of creation in six days has
nothing in common with the evolutionary account; even if each day
is arbitrarily interpreted as long ages, the sequence of appearance of
various living things remains incorrect. Indeed, young-earth creationists themselves often point out such problems in order to urge
that no compromise with evolutionary time scales is possible; hence,
the literal biblical account must be adopted by true believers. None of
the various holy writings of the world consistently get it right if they
are read as making claims about natural science.

Debates over the inerrancy of sacred books attract much popular attention, especially because not just nonbelievers but champions of

rival religions like to attack the veracity of what they see as fake scriptures. Within academic circles, however, such debates are usually ignored. The matter has been decided long ago: scriptures are, in fact, full of mistakes. Holy writings do not contain miraculous scientific knowledge or fulfilled prophecies. So in one sense, nonbelievers have already won over the intellectual world. Many fundamentalists certainly think so, regularly railing against the infidelity of the learned. Popular nonbelief, however, also gets little respect in academia. After all, sophisticated religious thinkers within the academic world are rarely fundamentalists; even when they uphold the Bible or Quran as divine revelation, they do not treat it as direct divine speech unaffected by the human fallibility of prophets and scripture writers. So nonbelievers making much of scriptural contradictions and scientific absurdities can come across as beating a dead horse. Worse, they are perceived as sharing the same narrow misunderstanding of spirituality as their literalist opponents.

Still, the realization that science and a literal reading of scripture conflict remains important for many nonbelievers. Many skeptics who grow up in a conservative religious environment come to explicit skepticism by first trying and failing to resolve the apparent contradictions and mistakes in holy writ. Fundamentalists justly worry about the risk of college students losing the faith into which they were born. Of those who drift away, most remain believers of some kind, adopting more liberal views. Some, however, become skeptical about religion, unless they find some other source of supernatural conviction drawing them to alternative spiritualities.

FROM SPIRITUALISM TO THE NEW AGE

Scriptural literalism does not answer everyone's religious needs. Another way popular religion adapts to modern times is to bring forth a more ecumenical brand of supernatural power, available to all. And just as many fundamentalists try to claim support from science, even forming scientific-seeming organizations to promote creationism, paranormal enthusiasts also try and enlist science in their cause.

Traditional religions easily accommodate movements emphasizing contact with the supernatural, such as Pentecostal sects driven by spirit-possession experiences. Similarly, magical healing is no stranger to old-time religion. In the nineteenth century, however, some religious experiments in the industrializing Western world drifted away from the mainstream of Christianity. Spiritualists, in particular, strained the limits of orthodoxy. They focused on life after death rather than doctrine, and they were inspired by trance mediums who communicated the words of spirits and by physical mediums who levitated, tipped tables, and produced other manifestations of spirit power (see Figure 5.1). Some scientists who were disillusioned with traditional religion but who also worried about the growing influence of materialism were sympathetic to the Spiritualist movement, hoping that by investigating mediumistic phenomena, they might verify that the spirit survives death (Oppenheim 1985).

Spiritualists thought they could *demonstrate* the reality of spirit, rather than take it on faith. Hence they hoped to heal the breach between science and religion. They looked to paranormal phenomena which, in commonsense terms, were easily understood as manifestations of a nonmaterial reality. Strange phenomena and mysterious experiences were, just as importantly, available for all to witness. Religious truth was not the property of either a class of religious scholars or a scientific elite. And spiritual experience was above all practical. Whether in communicating with a dead sister or healing disease, contact with the spirit realm produced tangible rewards, not pie in the sky. In the English-speaking countries, Evangelical Christians taking the Bible literally democratized traditional religion, especially appealing to the lower strata of society. In contrast, Spiritualists were more middle class, and mostly dabblers rather than fully committed believers. Their spirituality was also democratic, but their style was more individualistic.

Much the same can be said about today's individualist, alternative religious currents. New Age enthusiasts, devotees of alternative medical philosophies, and even many liberal Christians have a nondogmatic spiritual outlook permeated by a diffuse sense of a supernatural presence in the world. Curiously, although this is the religious style of postmodern consumers rather than land-bound peasants, New Age notions of supernatural power are very similar to what is found in religions such as folk Catholicism or Eastern Orthodox Christianity, Tibetan Buddhism, folk Islam—even the shamanistic religions of tribal people that so often fascinate New Agers. Miracle beliefs have a practical focus: miracles validate religious commitment, produce healing, and allow communion with a spirit world. Yet, there are differences. Today's spiritual consumer picks and chooses, concocting a highly in-

Figure 5.1 Daniel Dunglas Home, a famous nineteenth-century medium, convinced many observers that he was able to perform feats such as levitating. (Topham/The Image Works)

dividual mix from a large variety of spiritual beliefs on offer (Roof 1999; Melton and Lewis 1992).

The New Age also remains ambivalent about science. Like their Spiritualist forebears, New Age seekers conceive of themselves as spiritual investigators open to new experiences. They have also inherited, however, a suspicion of mainstream science, which stubbornly refuses to acknowledge their miracles. To New Agers, science, which has done so much to disenchant the world, appears to be as much in need of serious reform as is traditional religion. More important, because they take extraordinary experiences at face value, New Agers tend to distrust the theoretical side of science. While scientists seek knowledge of the constraints on reality, individualistic spirituality is more concerned with freedom. Scientists try to have theory and experiment correct one another, but in a New Age environment, accepting individual experiences at face value is the prevailing mood.

Science-inspired nonbelievers typically have no more use for the New Age than for scriptural literalism. Indeed, the New Age has a reputation for lacking seriousness. The paranormal is associated with tabloid psychics who predict Atlantis will rise from the ocean next year, hucksters who fleece sick people with the latest "alternative therapy," and astrological pronouncements that fit just about everyone who reads them. TV mediums like John Edward and superpsychics such as Uri Geller are the public faces of psychic power, and their images are tainted by suspicions that they take advantage of the gullible.

So the paranormal attracts the ire of skeptics, whether they go as far as denying all supernatural claims or if they are only interested in defending mainstream science. To the skeptics, growing popular interest in the paranormal often seems like a disturbing wave of irrationality that needs to be opposed for the greater social good (Kurtz 2001).

As ever, nonbelief is not simply the opposite of traditional religiosity or even its main competitor. Individualistic, occult spirituality is a third force to be reckoned with. And so the competition between belief and nonbelief involves shifting political alliances. Naturalistic nonbelievers share many interests with New Agers and liberal religious people when defending a modern, individualist social environment. But when opposing paranormal claims, naturalistic skeptics can easily find themselves joining forces with defenders of more traditional religions. Even fundamentalists, who scorn scientific opinion on matters such as evolution, readily cite scientific skeptics when they argue that astrology or transcendental meditation are not what New Agers advertise them to be.

SKEPTICS VERSUS PSYCHICS

The argument between skeptics and proponents of the paranormal largely takes place outside of academic journals. The popular case for paranormal miracles is typically built out of anecdotes: stories of how a predictive dream came true, the spectacular success of a spiritual healing episode, or how a psychic did something incredible on a TV show. Even more serious arguments that there is something beyond the natural world are often based on anecdotes, now expanded into case histories.

Finding a detailed natural explanation for every dream of a grandmother's death followed by a phone call announcing her passing would be an impossible task. So skeptics have to argue that in general, the sorts of stories that make up popular evidence of miracles cannot be taken at face value.

Believers take the low level of scientific interest in paranormal claims as a sign that science is being closed minded, or that paranormal phenomena fall outside the narrow scope of conventional scientific methods. Skeptics, on the other hand, say that most scientists ignore the paranormal because the evidence supporting miracles is of very low quality. Anecdotes, in particular, are not good evidence—the skeptics' slogan is "the plural of anecdote is not data." A story is just a story. It can be very hard for an investigator to figure out whether a strange tale leaves out crucial details, is accurate in its critical claims, or if it has been exaggerated. We have no control over anecdotes—they are not like laboratory experiments where scientists can isolate what might be relevant variables and repeat the test until they are confident they understand the phenomenon. Some especially strict skeptics bring up an argument that goes back to the Reformation, when some Protestants wanted to dismiss the miracle claims associated with Catholic saints. As refined by philosopher David Hume (1992), this argument suggests that no collection of miracle stories can carry more weight than the reliable and continually repeated evidence that the laws of nature are never violated. Applied to paranormal tales, the Humean argument says that even lots of anecdotes never add up to good evidence for the paranormal.

Discarding anecdotes altogether probably goes too far. After all, we still have to ask how miracle tales come about. Why do so many sane people, from different cultures and backgrounds, report paranormal phenomena? Why are their reports roughly similar—ghosts, precognitive dreams, feats of mind over matter? Now, skeptics are correct to insist that jumping to the conclusion that miracles happen is unwise. As another slogan goes, "extraordinary claims require extraordinarily strong evidence."[1] If an acquaintance tells us that fuel is unusually cheap at the local gas station, we can usually take her word for it. But if she reports space aliens landing in her backyard, we should want stronger evidence before accepting her claim. Still, we need some kind of explanation for how we end up with so many paranormal reports if there is no magic in the world.

Skeptics begin by pointing out that many paranormal claims are the result of fraud or hoaxes. Crop circles—elaborate patterns that appear on fields overnight—appear to be of this sort. Many crop circle makers have come forth or have been exposed. We know a great deal about their various techniques. So we do not need to find the perpetrator of every crop circle to figure out that probably they all are human made. Many true believers remain who continue to think there is something paranormal—perhaps alien—about crop circles. But the circles we know all fall within the range of the sort of thing done in hoaxes. Nothing stands out as extraordinary.

Likewise, physical mediumship and psychic feats are suspiciously similar to the sort of performances given by conjurers. The slate-writings allegedly produced by spirits, spoon bendings, divinations of concealed drawings or contents of sealed envelopes, and just about everything in the psychic repertoire are accomplished by trickery just as well. In fact, from Harry Houdini to James "The Amazing" Randi, many famous conjurers have publicly exposed the tricks of psychics and Spiritualists (see Figure 5.2). Today, Randi heads an educational foundation devoted to skepticism about paranormal claims, and offers a million-dollar-plus prize for any psychic who can demonstrate a miracle under controlled conditions.

Skeptics are especially disturbed by the fraud that takes place in magical healing. Desperate people, often terminally ill, seek hope in the practices of psychic surgeons, Laetrile purveyors, and "energy healers" of all sorts. Sufferers flock to faith healers, rising from their wheelchairs or throwing away their medications. And almost always, after the initial euphoria passes, the patients' state of health returns to its former state. If they abandon conventional medicine because of their belief in a supernatural cure, they can get severely harmed as well as lose money.

Fraud and hoaxes are certainly common. Even so, it will not do to suggest that fraud is the principal reason behind miracle claims. Skeptics, in fact, usually argue that self-deception is a more important contributor to paranormal beliefs than plain dishonesty. Even the most blatantly absurd forms of supernatural healing, for example, garner glowing testimonials to their effectiveness. The will to believe can be

Figure 5.2   Harry Houdini, the most famous magician of his day, exposing techniques used by fraudulent mediums on the stage of the New York Hippodrome, 1925. (Library of Congress)

powerful, and intentional fraud thrives on this psychology. James Randi once exposed a faith healer, Peter Popoff, who appeared to have uncanny knowledge of the condition of patients in his audience. Randi found this knowledge was available to Popoff by means of a confederate and a concealed radio receiver (Randi 1987). Though he was disgraced, some years later Popoff again set up shop with his healing ministry. His audience wanted to believe.

In fact, no one involved in paranormal practices need have any intent to deceive. Most psychics and healers are completely sincere. They are convinced that they are using a divine gift for the benefit of others. Many patients are satisfied by their experiences with healers and attribute any sense of feeling better to the healer's special powers. Their testimony further confirms the healer's belief in her own magical ability.

Similarly, when a psychic gives relationship advice to a client and the client readily believes in the psychic's paranormal powers, both parties can easily collaborate in self-deception. Ray Hyman, a psychologist and prominent critic of psychic claims, started out practicing palm reading and intuitively divining the personal problems of his clients. Without realizing it at the time, he was using techniques of "cold reading"—all the uncannily accurate knowledge he divined about his clients was derived from vague statements that clients interpreted to fit themselves and from feeding back to them specific information clients had already given to him. Though nothing paranormal was happening, psychic and client were constantly reinforcing each other's belief that something magical was taking place (Hyman 1996).

Skeptics also point out that our tendency toward magical thinking is reinforced by spooky natural events for which we might not know the real explanation. For example, a common New Age event is a fire-walking session where people quickly walk over a bed of hot coals. It can be hard to resist interpreting fire walking paranormally, as a confidence-building demonstration of mind over matter. Fire walking, however, is actually an illustration of basic physics: the hot coals conduct heat poorly and have a low heat capacity to begin with. The reason a fire walker's feet do not burn is closely related to why our hand will burn if we touch a cookie sheet in a hot oven, but not if we briefly stick our hand into the oven air that is at the same hot temperature.

Sometimes a knowledge of psychology helps demystify a spooky event. Many people have occasional experiences of waking up from sleep but not being able to move a muscle, often accompanied by a feeling of being oppressed. Traditionally, this was thought to be an attack by a demon or a night hag; today, many interpret the experience as part of an abduction by space aliens. Its causes are not, however, so mysterious. Especially when we are about to fall asleep or about to wake up, the mechanism for keeping our body immobile during sleep can occasionally work improperly. The resulting state often includes difficulty breathing and is very conducive to hallucinations (Zusne and Jones 1989). For those who experience it, the seeming attack by a night hag or abduction by an alien can seem very real and it is certainly very disturbing. Nevertheless, it has roots in a physiological malfunction, which is then overlaid by hallucinations and cultural elaborations of the event.

Some other aspects of human psychology that feed into paranormal beliefs are the difficulty we have in judging probabilities and coincidences, and the way we tend to notice instances that confirm a belief rather than those that do not. Dreaming of a grandmother's death the night before she dies is likely to be remembered much more vividly than those cases where the predictive dream happened but the grandmother remained alive. We should consider the frequency of all such dreams, whether the prediction was successful or not. And we should also consider the probability of someone, somewhere in the world dreaming that a relative will die. That impressive predictive dreams should happen by sheer chance is not that surprising.

To skeptics, the complexity of the natural world is also a reason to hesitate to proclaim that puzzling phenomena are miraculous. A completely natural world is still full of surprises. For example, in some lucky cases, cancer goes into spontaneous remission. Medical science is far from fully understanding cancer, but there is no reason to suspect anything supernatural. With something as complex as the human body and its environment, it is impossible to rule out uncommon instances of remissions that cannot be fully explained but are still due to natural causes. People who follow a magical healing procedure will also on occasion be spectacularly cured. Unless cures with such a procedure happen significantly more often than with conventional methods, there is no reason to believe that the healing procedure had any effect, let alone that it worked through magical means.

Similarly, there is the "placebo effect" in medicine, where patients respond positively to the mere belief that they are being treated even when they are given sugar pills. This is no great surprise, as a positive state of mind and reducing stress can strengthen the immune system and help recovery, but the details of how the placebo effect works are not well understood. Nevertheless, the effect is strong enough that in clinical tests of treatments, a control group has to be given a placebo to check if the treatment does anything over and above what comes from just the feeling of being treated. Magical healing methods are rarely tested this way, which makes it doubtful whether they have any real effect, let alone supernatural significance.

Skeptics argue that although we live in a world where miracle stories come at us from all directions, this is not because paranormal phenomena are real. We make mistakes, and we too easily interpret strange occurrences as supernatural events (Schick and Vaughn 2005).

MIRACLES IN THE LABORATORY

Especially from a scientific point of view, a major problem with miracle stories is the lack of control. In the nineteenth and early twentieth centuries, Spiritualist physical mediums worked in the near dark. Today, a psychic bending spoons does his wonders in everyday circumstances with all their distractions, not in an uncluttered room under constant videotaped observation from all angles. The performer calls the shots. Although defenders of the paranormal say that psychic powers are fickle and work best in a relaxed environment far from skeptical scrutiny, such conditions are also ideal for deception and misperception.

Early psychical researchers soon realized that collecting wondrous stories and reporting what happened in particularly impressive séances was not enough to convince scientific colleagues who were used to more solid evidence. In the twentieth century, psychical research entered the laboratory under the name of parapsychology. The quest for the supernatural would continue, but now under controlled conditions. Repeatable experiments similar to those used in the behavioral sciences would produce good evidence for the soul.

Laboratory parapsychology tries to reduce supernatural feats to their essentials. Instead of a collection of highly individual miracle stories full of the idiosyncrasies of different spiritual traditions, parapsychologists investigate the possibilities for mind reading, obtaining information about the world when no ordinary physical means are available, and manipulating matter by pure intent without any physical interaction. They use ordinary people as subjects, not just psychic superstars. And they put their subjects into spare laboratory environments, asking them to do simple tasks such as guess cards or attempt to influence a physical event just by thinking about it (see Figure 5.3). By repeating these tasks over and over, parapsychologists try to find a psychic signal. This signal would be visible as a statistical deviation from the chance-level performance expected in a world without magic. Thus, parapsychologists hope to present well-controlled, repeatable, and quantified evidence that there is more to the world than allowed for by materialists (Radin 1997).

Unfortunately, laboratory parapsychology has suffered from many of the same problems that plague psychical research outside controlled settings. Even fraud has all too regularly been discovered in important parapsychological work. The problem continues today. Recently, the public interest in magical healing has given parapsychology both a new direction of research and an important source of funding. A number of studies of prayer having an effect on healing or fertility have been conducted, some reporting that remote prayer done without the knowledge of patients results in measurable benefits. One of the most touted studies has now come under the strong suspicion of outright fraud, alongside other flaws (Flamm 2004).

Figure 5.3 Woman undergoing a ganzfeld experiment to test for telepathy, trying to receive other people's thoughts. Red light is shone through ping-pong balls on the eyes of a receiver while white noise is played through headphones. This sensory deprivation results in hallucinations. Several senders then try to transmit randomly chosen visual stimuli, which the receiver attempts to describe. Finally, the receiver has to select by sight the target from a selection of stimuli. (Jeremy Walker / Science Photo Library)

Other studies that find a healing effect of prayer unknown to the patient also suffer from numerous deficiencies. As the quest for miracles moves into the laboratory, however, critiques of paranormal claims become more technical, and finding errors in experimental protocols requires more expert scrutiny. For example, some prayer studies have had flaws in their blinding—ideally, neither the patients nor the evaluators of their health should know if they are being prayed for or not—and their statistical procedures. Subtle problems are not easily detected even during the peer review of the research; often skeptical objections surface after publication and after the media has already trumpeted the paper as scientific confirmation of the supernatural power of prayer.

Skeptics argue that flaws similar to those in the prayer studies are typical of the whole parapsychological enterprise. Such critics typically go hunting for mistakes in experimental design, such as allowing too many possibilities for normal sensory leakage, and often find them.

However, there is also the question of what to make of research that does not, on the face of it, appear to be flawed. Skeptics expect to come across some such research. After all, finding flaws in a research project is a difficult, painstaking task. Skeptics can no more guarantee they will find a nonmagical explanation for every experimental result than a detective can promise to solve every murder. Furthermore, experiments can report spurious positive results even if they are flawless. The criterion for statistical significance in parapsychology allows, as in the behavioral sciences, a 1 in 20 chance of the result being due to chance alone. This brings up the "file drawer" problem. It becomes impossible to judge what a published positive result means without having an idea of how many studies were not published and left in a file drawer because a negative result was not interesting.

The deeper problem with the laboratory evidence for psychic anomalies is, however, that none of the feats claimed are impressive. They do not stand out from among the mistakes and spurious results that occur in a nonmagical but complicated world. Some of the best evidence for miracles today consists of results such as changing the output of a random number generator once in every 10,000 tries. Where such marginal effects are concerned, the possible sources of error proliferate out of control—no one can reasonably expect to pinpoint the flaw in the experiment that might have led to such a minute deviation from chance. Our world is simply too noisy for researchers to reliably extract such a small psychic signal, even with the best-designed experiments. To establish that something beyond ordinary physics happens in a parapsychology laboratory, we need more than just a well-controlled experiment. We need repeatable results that are truly spectacular. Marginal deviations from chance are worthless as evidence (Edis 2002, chapter 6).

Skeptics looking back at the evidence presented by parapsychologists conclude that there is likely no psychic signal hiding in the noise. After all, reducing the noise by doing better controlled laboratory experiments seems to eliminate the more spectacular feats that spur belief in the first place. The miracles on offer become ever more modest as the investigation tightens and the possibilities for mistakes diminish.

Psychical research started in the nineteenth century, and its history gives no reason for confidence. After well over a century of effort, there is nothing that stands out as compelling evidence for a miracle. Parapsychology always has a set of claims on offer, which are supposed to include *the* repeatable experiment, *the* strong signal of magic. In time, however, critics start whittling away at the claims, and parapsychologists drop them and move on to another set of experiments. Many skeptics argue that this dismal history is good reason to believe that the claim of psychic wonders has been thoroughly tested and found wanting.

UFOS AND ALIENS

Traditional and psychic miracle claims are not the only challenge to the mainstream scientific picture of the world. Skeptics argue not just against parapsychology and creationism but against a host of claims on the borderland of science such as UFOlogy and claims of alien visitations.

Although ideas about space aliens and reports of strange aerial phenomena had already been around, interest in flying saucers took a firm hold in popular culture after the Second World War. As the Cold War between the United States and the Soviet Union began, unidentified flying objects spurred speculation about newly developed aircraft or even spaceships visiting from much farther away. It soon became evident that most sightings were misidentifications of natural phenomena: the planet Venus generated many UFO reports, as did meteors, odd cloud formations, and weather balloons. Not all sightings were easily explained, however, and some UFO investigators thought this residue of tough cases had causes beyond the ordinary. The more spectacular cases looked like interactions with intelligently guided craft, or produced witnesses who reported glimpsing the inhabitants of flying saucers.

The idea of alien visitors caught on with the public. A handful of scientists also came to think there was something to all the flying saucer reports, but UFOlogy developed as a kind of folk science, relying on enthusiasts to gather sightings and propose explanations. UFO reports were varied, from basic strange lights in the sky to encounters that looked like they had to be real alien contact if not hoaxes. "Contactees" appeared who told of communicating with "space brothers," even being invited onto their spacecraft for a tour of the planets. A cottage industry of flying saucer photographs sprung up. A UFO mythology developed and matured; the iconography and stereotypical story lines concerning aliens became known by almost everyone in our media-saturated societies.

Though immensely popular, UFOs never found favor in respectable circles. UFOlogists wanted to get mainstream science to acknowledge UFOs as a genuine mystery worth serious investigation, maybe even to recognize the reality of visiting space aliens. A collection of anecdotes about flying saucer sightings did not, however, impress many scientists, especially when they were the sort of thing that could easily be due to perceptual mistakes or hoaxes. So starting in the 1970s and 1980s, UFO claims steadily became more spectacular, with crashed saucers and alien abductions taking over the limelight (see Peebles 1994 for an overview of UFO history).

A flying saucer crash, such as the one UFOlogists have claimed happened in Roswell, New Mexico, in 1947, would be perfect for UFOlogy, which had never been able to produce convincing physical evidence. With a crashed saucer, UFO enthusiasts could point to a wealth of physical evidence, including craft and alien remains. However, the Roswell story also became entangled with the conspiracy thinking that had been part of the UFO subculture since its earliest Cold War days. The physical evidence, UFOlogists said, was real but it could not be produced because it was covered up by the government. So instead of yielding physical evidence, Roswell became another story based on anecdotes from people who claimed to have seen the evidence. In the 1990s, it became clear that part of the Roswell story was based on the crash of a secret military balloon, known as Project Mogul, which had come down near Roswell in 1947 (Saler, Ziegler, and Moore 1997). Skeptics were satisfied that the minor mystery of Roswell had been solved, but many UFO enthusiasts remained believers. To them, the Mogul story was yet another layer of the cover-up, and the alien bodies and technology still remained hidden somewhere.

Alien abductions, the other recent UFO concern, moved UFOlogy away from its "nuts and bolts" interest in physical spaceships toward a more spiritual direction. Popular UFO belief had always had supernatural overtones, and some small religious groups had taken

UFOs to be spiritually advanced saviors from the sky. Abductions provided UFOlogists with a spooky experience that was supposed to be direct contact with otherworldly beings. The abduction experience, however, was not just a relatively ordinary experience of being taken aboard an advanced spacecraft. We now have a standard narrative of experiencers being abducted into a UFO, typically at night and from their beds in a paralyzed state, encountering alien creatures and being drawn up to the craft (see Figure 5.4), even passing through walls, being probed and examined in uncomfortable ways, and finally being returned to bed and induced to forget everything. Believers in abductions say that although it seems forgotten on the surface, the traumatic experience manifests itself in psychological disturbances. Typically, a therapist who believes in alien abductions will suspect that a patient suffers from abduction-related problems and then attempt to recover the buried memories under hypnosis. Often the process takes place in a New Age environment, and believers find spiritual significance even in the terrifying aspects of the abduction.

Today, UFOs are spiritually ambiguous, appealing to both religious believers and nonbelievers. Aliens have become part of the New Age cultural landscape. More traditional religions can also make use of UFOs, however. Consider near-death experiences (NDEs), where a person who is near death feels as if her soul detaches from her body, passes through a tunnel of light, and meets dead relatives and religious figures before returning to life (Bailey and Yates 1996). Both parapsychologists and some conservative Christian writers take NDEs to be direct confirmation of a soul surviving death. The Christians, however, interpret NDEs in the context of their own worldview, emphasizing that the impression given by New Age NDE writers that every dead soul goes to a heavenly place must be a satanic deception. Similarly, many writers fit stories of UFOs and abductions into a traditional religious framework. Instead of aliens, they speak of angels and demons as being responsible for UFO phenomena. Strange phenomena are supposed to validate the supernatural through direct experience, and many people look to long-standing religious tradition to make sense of it all.

There is even a nonbelieving streak in UFOlogy. If some Christians interpret UFOs as angelic or demonic manifestations, the flip side is that some religious dissidents say the biblical stories are really distorted recollections of interactions with space aliens. Some UFO literature claims that Ezekiel's wheel (Ezek. 1:5–23 and 10:2–21) was an

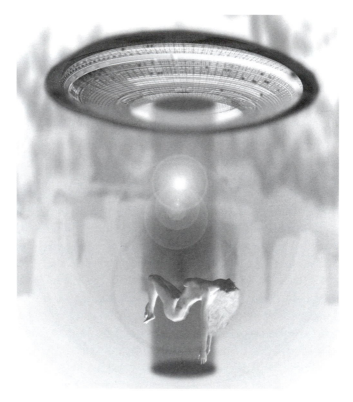

Figure 5.4 Computer artwork of a woman being abducted by aliens in a flying saucer. (Victor Habbick Visions/Science Photo Library)

alien spaceship (see Figure 5.5); others say that angels and even the God who set up the Garden of Eden were aliens. The Raëlian quasi-religious movement claims the human race was created by alien genetic engineers, the Elohim. The Raëlians promise to unite science and religion by replacing supernatural miracles with the actions of superior but merely natural intelligences. Science-minded nonbelievers have no more use for the Raëlian movement than for scriptural fundamentalism, but such UFO sects are as much a rationalist heresy as a Christian deviation.

Another connection between UFOs and nonbelief is the very idea of intelligent life on other worlds. Conservative Christians have been inclined to think humanity is unique in the universe; after all, they believe God became human. Intelligent design advocates today continue

Figure 5.5 Some UFO enthusiasts interpret Ezekiel's vision, "And I looked, and behold, a whirlwind came out of the north, a great cloud, and a fire infolding itself, and the likeness of four living creatures," as an encounter with an alien spacecraft. (From Royaumont, *Histoire de l'Ancien et Noveau Testament* [1724], p. 311. Mary Evans Picture Library).

to argue that the earth is specially placed to allow life, unlike elsewhere in a largely hostile universe (Gonzalez and Richards 2004). And Western nonbelievers, reacting against Christianity, have been eager to think intelligent life must abound in the universe. Currently, astrobiologists lean toward thinking that simple life such as bacteria may well be common in our universe, but that questions about the abundance of intelligent, technological life are much more wide open (Goldsmith and Owen 2001). The issue is of minor significance, though, since no one has a clear idea how abundant life should be either in the presence or the absence of supernatural powers.

DIRECT EXPERIENCE OF THE PARANORMAL?

Though UFO beliefs can be religiously ambiguous, many UFO enthusiasts think of alien encounters as paranormal phenomena that challenge a materialist picture of reality. So science-minded nonbelievers join with skeptics of the paranormal to defend mainstream science against the more extravagant UFO claims.

The flaws skeptics find in UFO claims are very similar to the problems with parapsychology. The cases for aliens and for psychic power both depend heavily on testimonials and are weak on hard physical evidence. Both point to a residue of unexplained cases, but this residue does not stand out from among the sort of difficult unsolved cases we can expect in a complicated world. Both fields are plagued by hoaxes and are vulnerable to self-deception. Moreover, trying to find unexplained anomalies means that both UFOlogy and parapsychology define their phenomena negatively—as something lacking an explanation—rather than present a well-developed alternative explanation. In other words, paranormalists do not have any theory to frame their investigations, and from a scientific point of view this is a serious handicap.

So, the interaction between UFO skeptics and proponents is similar to the situation with psychic power. The leading UFO claims change over time, as mainstream science keeps responding skeptically. Today, reports of strange sights in the sky have diminished, and UFO enthusiasts' interest has shifted to abductions and crashed saucers. Even many UFOlogists think their field is veering off course. Karl T. Pflock suggests that aliens had been visiting us in the 1940s and 1950s, but they have now left (Moseley and Pflock 2002).

But what about direct experiences of aliens, as related by abductees? Or, for that matter, the direct experience of an afterlife related in NDE accounts?

Some people with a scientific background do think UFO abductions are real paranormal experiences. Psychiatrist John Mack (1994) not only thinks they are real, but argues that UFO abductions present a serious challenge to the materialism of modern science. Skeptical psychologists are less impressed. A major difficulty with abduction reports is that they are typically produced under hypnosis. Hypnosis is popularly thought to induce a psychological state in which people can be commanded to do strange things. Hypnotized subjects are also supposed to recall events with photographic detail, even if the memory has been repressed and unavailable to their ordinary consciousness. These popular notions are not true. Hypnosis does not improve memory. In conditions of heightened suggestibility, hypnotized people confabulate, making up stories to fit the expectations of questioners. Unintentionally, therapists can easily lead some of their hypnotized subjects into constructing an abduction story. In fact, some psychologists argue that hypnosis, UFO abductions, multiple personalities, and even the spirit-possession experiences central to many re-ligions are related phenomena, and that they are largely social per-formances rather than extraordinary psychological states (Spanos 1996).

Other psychological factors also contribute to abduction reports; for example, the kinds of hallucinations that occur when a person is close to falling asleep or waking up, which lead to night hag types of ex-periences, seem to be a frequent element in abduction stories. Still, while UFO abductions are not direct experiences of the paranormal, neither are they pure examples of anomalous psychology.[2] An abduc-tion story is a cultural creation: strange, frightening experiences are shaped, elaborated, and given meaning in a shared cultural context. Stereotyped abduction scenarios are instrumental in interpreting the abduction experience, and the story of each individual abduction feeds back into the UFO subculture to strengthen or modify the stan-dard abduction scenarios.

Much the same can be said for other paranormal experiences such as NDEs. Mainstream scientists have a good, although still incom-plete, idea about the physiological basis for near-death experiences. Recognizing how the brain can adopt the hovering-above perspective of spatial memory as the most stable model of reality when cut off from external input helps explain the feeling of floating above the body much better than does the notion of a soul departing the body (Blackmore 1993). Medical details of the dying process, such as oxy-gen starvation, do much to illuminate features such as the "tunnel of light" that the typical individual enters (Woerlee 2005).

Again, however, the social aspects of how near-death narratives are constructed are just as important (Fox 2003; Kellehear 1996). There is no single standard "near-death experience," and no single cause can account for every NDE. An NDE is often life changing, and it is very easily interpreted as direct contact with a spiritual realm after death. Nevertheless, it is not a direct contact with a supernatural reality. NDEs are generated through a complex interaction of culture and our built-in tendencies because of the structure of our bodies and brains. They undeniably feel real, but they can be explained within the natu-ral world.

Many modern religious movements emphasize personal experience of the supernatural. When scientists express skepticism, they can come across as ignoring direct experience, favoring the abstract theo-ries of modern science over the plain facts available to anyone open to the divine. Even so, science cannot avoid skepticism. Direct experiences infallibly telling us about the way the world works are simply not available. A theoretical framework, whether explicitly elaborated or implicit in our culture or brain structure, is always essential for getting a handle on the world. This is not to say that scientists must disregard all strange reports unless they originate from a lab, are recorded by instruments, or are captured by theoretical explanations. Testimonials and claims of direct paranormal experience do not, however, count for much. They give us a starting point for doing science, not incontrovertible facts that must be accepted by all.

So, miracle claims continue to inspire doubt in the culture of science. Not only the psychics and healers of today, but historical testimonies about miracles are also suspect—whether they are about Jesus, famous rabbis, or Buddhist teachers. Naturally, the kind of nonbelief inspired by science usually includes a strong skepticism about miracle stories, or indeed about any claim to raw, unmediated experience of the nature of reality.

PARANORMAL SKEPTICISM AND NONBELIEF

In scientific circles, the paranormal has a bad reputation. Still, since magical beliefs are so popular, scientific institutions come under pressure to respond.

To many scientists, the fact that more people are fascinated by UFOs or psychic powers than by genuine developments in science indicates a basic lack of public science literacy. If only science education worked better, they hope, bookstores might no longer have larger sections devoted to astrology than to astronomy. Science-minded nonbelievers echo such concerns. Moreover, many nonbelievers suspect that the popularity of supernatural belief of all sorts is largely due to the credulousness and even outright ignorance of most people. They think that if our cultures did not insist on supporting social order with supernatural doctrines—exalting faith over reason—at least fundamentalist religion and the more frivolous New Age fads would die down.

While most scientists and nonbelievers would like to see more skepticism about fringe science and the paranormal, this clearly is not happening. In fact, the more popular fringe-scientific beliefs generate no end of political headaches for the scientific community. In the United States, with its conservative religious culture, creationism and intelligent design continually threaten science education. A more recent development is how parapsychology and alternative medicine have made great strides toward broad public acceptance in even the most technologically advanced countries. Magical healing does not impress natural scientists and defenders of mainstream medicine, but large numbers of customers spending billions of dollars every year means a good deal of political clout and many institutions wanting a share of the available money (Sampson and Vaughn 2000). Today, many insurance companies pay for alternative therapies such as homeopathy regardless of their scientific absurdity and complete lack of good evidence of their efficacy. With the recent establishment of the National Center for Complementary and Alternative Medicine (NCCAM) as part of the National Institutes of Health, the mainstreaming of magical healing has reached a new level in the United States. Among other activities, NCCAM funds attempts at research by advocates of alternative treatments. Parapsychologists now hope to receive a steady stream of federal grants, as long as they propose to study psychic healing.

As unsubstantiated claims receive significant backing, skeptics and defenders of mainstream science enter the fray. Politically charged debates, however, do not allow extended discussion of scientific matters. Skeptics need to present sharp, easily understood distinctions between mainstream science and its competitors. They then have to argue that only mainstream science deserves public trust, without offending commonly held religious beliefs.

The easiest way to exclude the rivals of mainstream science is to call them "pseudoscience" and suggest that they are essentially different from real science. So skeptics and concerned scientists have long sought a philosophical test to distinguish science from its pretenders. For example, many skeptics have accused creationism, parapsychology, astrology, and so forth of presenting unfalsifiable claims (see Grim 1990). Paranormal claims, they say, cannot be put to the test and therefore cannot be examined by science. It is hard to sustain this view, however, when so much skeptical criticism of paranormal and fringe-science claims tries to show that these claims are false. The claim that Earth is only thousands of years old is false, not unfalsifiable. Certainly, paranormalists often respond to criticism by making excuses for failure, such as saying that skeptical investigators cannot reproduce psychic effects in the lab since skepticism inhibits psychic powers. Even so, this is not a fault intrinsic to the idea that psychic powers exist, only evasive behavior by some parapsychologists.

Philosophers of science have long tried to find some essential features of science, like falsifiability, methodological naturalism, and so on, which would distinguish science from pseudoscience. None of their proposals have really worked. Currently, most philosophers have abandoned the attempt to rigidly define science. Though many ideas on the fringes of science seem mistaken, creationism, parapsychology, and UFOlogy go wrong in different as well as similar ways; they do not fail a common test. If the label "pseudoscience" has any use beyond political rhetoric, it is as a description of an enterprise that claims to be advancing knowledge but is not structured to do so. The problem with creation-science, for example, is the way the community of creationists is structured to defend a predetermined religious view no matter what. There is nothing intrinsically unscientific in the hypothesis that biological species were separately created a few thousand years ago; it just happens to be almost certainly false according to the evidence we have.

A more nuanced critique of the fringes of science might make more intellectual sense, but it does not make much headway in a sound-bite-driven media culture. Worse, abandoning the effort to draw sharp philosophical lines between science and its pretenders goes against the liberal inclination to confine science and religion to separate spheres of truth. Saying "the Quran may well be true, it just is not science" is much safer than arguing that according to modern knowledge, ancient scriptures simply get it wrong.

As always, associating science with nonbelief is politically risky, particularly in conservatively religious countries such as the United States. Thus, skeptics about the paranormal have to be careful to know where to stop. They continually face the question: Should skeptics be content to defend mainstream science against challenges from the fringe, or should they extend their critique of the paranormal to more respectable supernatural beliefs?

The United States supports a number of skeptical organizations. The Committee for the Scientific Investigation of Claims of the Paranormal (CSICOP), composed largely of scientists and other academics, promotes mainstream scientific responses to paranormal claims, trying to represent the skeptical view to a wide audience. Other groups such as the Skeptics Society have a similar mission, including acting as sources of information for the media and publishing skeptical periodicals accessible to scientifically literate members of the general public.

Organizations such as CSICOP and the Skeptics Society emphasize critiques of paranormal and fringe-science claims rather than critiques of mainstream religious beliefs. They draw support from liberal religious people as well as scientists and nonbelievers. Nevertheless, they regularly get entangled with religious questions as well. Skeptics of creationism, aliens, and psychic powers cannot merely stand back and declare "insufficient evidence" all the time: they often have to present and defend an overall theoretical framework according to which psychic powers or intelligent design in biology should not be true. Where fringe claims are concerned, skeptics take a naturalistic view, and it is no great strain to extend this naturalism to religion in general (Edis 2004a).

Defenders of paranormal claims are quick to notice the nonbelieving streak within skeptical organizations and often accuse skeptics of being blinded by materialistic ideologies. Certainly the leadership of skeptical groups leans toward nonbelief. Paul Kurtz, chair of CSICOP, also heads the Council for Secular Humanism, devoted to presenting a comprehensive, naturalistic alternative to religion. Michael Shermer, who runs the Skeptics Society, is also a noted spokesman for agnosticism. This is not to say that there are no prominent skeptics who are religious, or that paranormal skepticism invariably leads to nonbelief. Nevertheless, there is a connection. In many ways, skeptics about the paranormal tend to represent the science-inspired style of nonbelief.

Historically, denying miracles was a secondary part of the philosophical tradition of doubt, taking a back seat to more grandiose metaphysical disputes. In modern times, as science set the standard for reliable knowledge, paranormal skepticism has become more important in the overall arguments for nonbelief. Skeptics can still scoff at New Age beliefs and denounce creationist anti-intellectualism while at the same time being content to leave more socially respectable supernatural beliefs alone. Still, skepticism about the disreputable fringe very often carries over into a full-blown naturalism.

NOTES

1. More usually, "extraordinary claims require extraordinary evidence," but the meaning of this shorter version is less clear.

2. Anomalous does not mean mentally ill. Abduction experiencers seem no different than the general population in terms of overall mental health.

Chapter 6

Explaining Religion

TOO MANY RELIGIONS

Why are there so many religions? Why do people believe so many incompatible things when it comes to spiritual matters? Anyone, skeptic or believer, who is confronted with the diversity of religious opinions faces questions. One way or another, vast numbers of people must be mistaken. But how do they come to make the particular mistakes they make?

Monotheists usually think revelation comes to distinct times and peoples. If others have not been saved, God must have a reason. Maybe the role of the faithful is to evangelize the heathen and bring them to Truth. But then, what about those who hear the divine message and still follow other gods or none at all? Many monotheists suspect that incorrect belief must be rooted in a moral failing. Pride, perhaps, misleads the obstinate infidel. Or they remain chained by those sins that give them pleasure.

More liberal-minded theists are willing to grant almost everyone a degree of divine illumination. All are touched by a hint of revelation—except, perhaps, those unfortunates unable to perceive any spiritual depth in the world. New Agers, who also emphasize tolerance, usually see diversity as just the way cosmic consciousness manifests itself. We should follow our own inspiration and not worry about individual differences of belief. Evangelizing monotheists arrogantly claim to possess a unique truth; they restrict the liberty of the spiritual seeker. Skeptics, in their own way, are also closed minded—they fail to trust spiritual experience.

Nonbelievers also face vexing questions. Why do the majority of people continue to hold beliefs that seem superstitious at best? Some nonbelievers think it is because "people are stupid" (Joshi 2003, 12–16), ignorant, or deceived by powerful religious establishments. Less crudely, nonbelievers have often thought that with better education, freedom from theocracies, and increasing mastery over our fates, ordinary people would come to see that the gods were not necessary for their lives. Religion exists because most of us do not know any better.

It appears we very often attempt to explain opposing beliefs as being due to personal failures. And the faults we see in others are usually those we worry most about. Monotheists anxious to obey their God often think of infidelity as sin, rebellion, or at least a self-centered lack of moral maturity. Postmodern New Agers see interference with individual liberty lurking behind any attempt to discover a truth beyond personal opinion. And nonbelievers, who think of themselves as defenders of reason, take spiritual beliefs to be caused by a failure of rationality.

Such proposals, besides being suspiciously self-serving, are not good explanations. There are far too many morally sensitive infidels, tolerant critics, and smart devout people in the world. The underpinnings of both belief and nonbelief have to be considerably subtler than a list of personal faults.

Science—including the social and behavioral sciences—should help us do better. Nonbelievers in particular have thought that we should be able to find a compelling scientific account of religion. If supernatural perceptions and religious behavior could be explained within the mundane world, a naturalistic view of the world would be vindicated by social as well as natural science.

Early social scientific attempts to explain religion often took Western culture as the norm, treating others as deviant. For a Christian anthropologist, "primitive" religions could be anything from barbarous superstitions to honorable yet still defective stages on the way to a mature Protestant commitment. The less devout thought that religions evolved from "animistic" beliefs to the grand metaphysical systems of the world religions, and all were finally superceded by positive science (Olson 2004, chapter 6).

Many observers also have thought that religion could be explained by figuring out what functions it performs for believers and for societies. Religion provides an answer to mysteries such as why the universe is as it is—or, alternatively, religion is anti-intellectual, preventing too many questions. Religion acts to keep societies together, allowing the devout to extend their family affections to the community of believers. Religion helps people overcome the fear of death. Or religion is a device of the ruling classes, the opiate of the people—or sometimes it is the sole voice of the downtrodden.

Like any social phenomenon, religion is complicated, so it is no surprise if many such explanations appear to have the ring of truth, or at least partial truth some of the time. Still, functional and progressive evolutionary explanations typically have been little more than informed speculation. Such approaches tend to ignore religious claims to present concrete truths about genuine supernatural realities. For the truly devout, a supernatural belief is more than sufficiently explained by an account such as, I believe because the Holy Spirit opened my eyes to His Truth. Only when such claims are no longer convincing does religion become a phenomenon to be explained from the outside.

Social science started out as an Enlightenment enterprise, infused with the anticlerical social ideals of the European Enlightenment. Even today, social scientists are more likely to be skeptical about conservative religious doctrines than natural scientists (Stark and Finke 2000, 53–54).[1] Nevertheless, today's social scientists are also more careful about setting aside questions about the truth of supernatural claims and instead examining the clearly social aspects of religion. In just about all cultures, there is widespread belief in supernatural agents who usually are closely involved with moral matters. The devout in all lands perform elaborate rituals that can seem bizarre to outsiders. Believers make costly sacrifices in connection with their spiritual commitments and receive no immediate tangible benefit. Such common elements of human behavior should be at least partly explainable by social science, regardless of whether any religion's description of the spirit world is accurate or not. If the explanations scientists come up with are any good, they may well have implications for the truth or falsity of supernatural claims. But there is no need to prejudge the question.

Today's social science also attempts to avoid treating religion as a peculiar feature of human societies in need of special explanation. In particular, from a social and psychological point of view, science and religion—or belief and nonbelief—should not have radically different causes. After all, in different ways, religious and scientific claims can both appear counterintuitive. The Christian belief that a virgin gave birth and became the mother of no less than the creator of the universe does not immediately square with common sense. This is not to say that religious and scientific claims are completely similar. Scientists studying religion have to ask what it is about supernatural agents that they so effectively grab human attention and put down deep social roots—unlike the much more recent, much less naturally appealing concepts of modern science. Nevertheless, when some people do not believe, perhaps preferring a scientific approach, this also calls for an explanation. Science-minded nonbelief must also have social causes and psychological mechanisms no less than religious belief.

RELIGIOUS STUDIES

Few would deny that social science can say some useful things about religion. After all, churches regularly commission sociological studies to find out membership trends and clues about ways to strengthen religious influence in society. Many students of religion, however, resist the notion that what really matters about religion can be explained in either social or natural scientific terms. They say that no outsider's description can ever capture what it is like to be religious—that it must be lived.

This resistance to science is sometimes expressed in a way that appeals to the academic divide between the humanities and the sciences. The humanities are not hotbeds of conservative religiosity—the kind of literal-minded fundamentalism that has trouble with ambiguity and metaphor in texts is not very appealing from a literary intellectual standpoint. Still, natural science aspires to be literal-minded in its own way, and thinkers in the humanities have a tradition of suspecting that science, however brilliant in cutting apart and modeling inanimate nature, overlooks the true depths in life. Human experience—the life of the mind in particular—seems far removed from the mindless interactions that are the domain of physical science. Natural science strives to be value free, but the questions that the humanities are interested in are always colored with moral concerns and commitments. Meaning is central to the humanities.

So while humanities-based intellectuals are often less religiously conservative, their culture is more hospitable to a kind of liberal transcendentalism. In this regard, social and behavioral science stands in between the humanities and natural science. Some, particularly psy-

chologists, are closer to natural science, seeking to fit human experience within the natural order. Others, such as many cultural anthropologists, conceive of their task as seeking interpretation and
sympathetic understanding rather than providing explanations.
Recently, postmodern philosophical currents that emphasize
humanities-style interpretation and deny the possibility of objective
scientific knowledge have become popular in the social sciences and
humanities, though they have made very little headway in natural science (Gross, Levitt, and Lewis 1996).

In today's academic environment, the study of religion takes place
in many departments, but the field of religious studies makes the most
direct claim to religion as its territory. And much in religious studies
consists of solid historical and textual research relating to various traditions from around the world. Still, religious studies more usually
takes the approach of the humanities and the more interpretive among
the social sciences. As a discipline, religious studies does not endorse
the specific doctrines of any one sect. Still, it is largely a liberal, ecumenical theological enterprise. Prominent practitioners of religious
studies insist religion is sui generis—not explainable on any but its
own terms. Their goal is understanding, not explanation, and understanding can ultimately be done only from the inside.

Much of religious studies, then, acts to protect religion from outside
criticism, especially in the United States (McCutcheon 2001; Wiebe
1999). In particular, it limits the possibilities for scientific explanations
of religion. This is not an explicit challenge to science like religiously
conservative attempts at creation-science and so forth. Instead, more
liberal, humanities-influenced intellectual subcultures demand that
natural science should stick to its own narrow sphere. Theological liberals are likely to claim there are "other ways of knowing" that are
just as valid in their own domain. They are convinced that where truly
deep and important existential questions are concerned, science has
very little to offer. Instead, we have to rely on an essentially religious
way of understanding.

In contrast, those thinkers inspired by science have a much more
expansive view of science. This is not to say that everything can be
taken apart under a microscope, or that the methods of physics provide the right approach to explain everything. Nevertheless, today's
heirs of the European Enlightenment continue to think that different
domains of inquiry are continuous with one another. They believe in
the overall unity of science. Hence, they like to push on with the ef-

fort to explain religion within science, rather than take it as an area immune to scientific investigation.

The argument over explanation versus understanding in religion closely resembles the debate over materialist approaches to the mind. This similarity is not surprising, as the common view in religious studies is implicitly dualistic, concerned to protect mind and meaning from the evil fate of being reduced to social and physical goings-on. The claim that no explanation can capture what it is like to be religious is equivalent to arguments that no materialist theory of mind can account for qualia. And defenders of scientific explanations of religion make familiar replies: no explanation captures firsthand experience, but that is beside the point. A simple change of perspective, itself no mystery from a materialist point of view, cannot set religion beyond explanation (Pyysiäinen 2004).

In that case, a scientific study of religion can proceed, with no apologies. Indeed, theological attempts to insulate religion from outside criticism are not so much barriers to investigation as an aspect of religion that is itself in need of explanation. Nonbelievers are liable to suspect that the sui generis "truth" spoken of in religious studies is indistinguishable from delusion, and that such truth would not be taken seriously if not for the fact that religion had begun to look false when subjected to modern criticism. Evoking paradox and making claims that are very difficult to test, however, are common features in religions, not confined to theological encounters with science. How is it that extravagant and hard-to-confirm claims, such as an afterlife or a realm of gods and demons, come to be deeply believed, are socially very common, and flourish even though they do not seem to confer any direct worldly advantage on the devout?

RATIONAL CHOICE THEORY

Sociologists of religion attempt to explain the human side of religion. Whether the supernatural claims of religions are true or not, what believers take to be contact with transcendent realities is always mediated through congregations, churches, saints, and so on. Sociology is a relatively young science, but we do know something about human social structures, and we can ask how social forces are organized in a religious context.

One prominent recent example of a sociological approach is rational choice theory, which applies a free-market economic model to religion. Rational choice theorists treat religious people as customers demanding spiritual services, rationally choosing what best meets their needs from among different religious organizations acting as suppliers. In that case, the religious landscape of a modern society would be best understood as a market for spirituality, shaped by supply and demand.

Religious demand, rational choice theorists believe, is determined largely by human nature. One influential account describes the service religion provides as compensating for the imperfections and suffering of earthly life (Bainbridge 1997, chapter 1). Almost everyone, for example, would like to avoid death—anyone selling an immortality potion would find a huge demand for her product. No such potion is available, but the demand for immortality still exists. In such a case, people will be attracted to an offer that postpones the reward of immortality, substituting a promise of an afterlife, which cannot be unambiguously put to an empirical test. Religions provide compensation based on a supernatural perception of the world. And since the imperfections of life never vanish—even though many of us are richer and healthier than our ancestors, we still die—the demand for religious compensators should remain roughly constant.

If the demand side of the religious market is relatively inflexible, the differences between the religious landscapes of modern societies must be due to the supply side. Some countries are more publicly religious than others, since demand can be suppressed or simply not met if there is a failure of religious supply. Religious people are rational economic actors; if the religious options on offer are artificially limited or are too high in cost, consumers will try and take their business elsewhere or get by without religion.

Rational choice theorists therefore expect that the religiously most vital societies are those with a free market in religion. If there are many different sects and varied styles of religiosity available to spiritual consumers, they will be more likely to find a religion that meets their demand. If different religious providers compete for religious customers, they will be motivated to offer a more attractive religious product, again contributing to the overall vitality of religion in that society. If, on the other hand, the freedom of the religious market is restricted, rational choice theory predicts religious participation will decrease. In European countries with state churches, the single approved religion on offer does not meet the broadest possible demand. Moreover, since the established clergies do not have to compete for souls for their livelihood, they get lazy; they are less motivated to make their product more attractive. In contrast, the United States, with its free market in religion, is much more religiously innovative, and its population has a much higher rate of religious involvement.

A rational choice approach is attractive because, inspired by economics, it promises a robust theory: something more than a description of religion in various societies, combined with speculation on social forces shaping religion. It is supposed to produce definite predictions about religious participation, which can be tested by empirical data such as survey results. Rational choice theory is also attractive because it does not ask the question, Why are people religious? with the implication that being religious is puzzling because it is a mistake made by otherwise reasonable people. Religious behavior becomes no different than the normal, rational choices people make in other aspects of their lives. In fact, rational choice theorists answer the question that frustrated nonbelievers ask, Why there is so much religion left in an age of science? by arguing that science can provide no intellectual criticism of religion. They support this contention by survey data showing that large numbers of scientists remain conventionally religious (Stark and Finke 2000, chapter 2).

Human societies are complicated, and so every sociological theory has limitations. Rational choice is a theory of modern religion, in societies that can support a large religious market, and where individual consumers can shop for faith rather than almost always stick with the religion into which they were born. In fact, rational choice theory is at its most plausible when describing a North American, Protestant-style religious marketplace. Most of the world lives according to very different forms of religiosity. Few can accuse the Islamic world of lacking religious fervor, but the religious organization of Muslim societies is hard to describe in terms of competing religious suppliers. Many of the more religious societies in the world are dominated by one religion that actively limits choice, such as Catholic Ireland and Poland.

Critics point out such shortcomings and go on to say that rational choice theory suffers from many other problems as well. The theoretical apparatus borrowed from economics looks attractive, but this also means importing the difficulties of economic theories. Real economic activity takes place in a world of imperfect information, faulty rationality, market failures, and pervasive breakdowns of free-market assumptions. Notably, the instrumental rationality of a pure consumer works well in harsh, constrained environments, but is too narrow and blinkered otherwise. Perhaps most damning, however, are the empirical failures of rational-choice-based predictions in religion. Even in the United States, where the theory might be expected to apply best, its predictions fail unless Catholics are treated as exceptions (Bruce 1999). Rational choice theory seems to fit Protestant-style religion, and even that is true only in a limited range of circumstances. This is hardly the mark of a theory that can claim deep insights into the nature of religion.

Another source of worry is the ideological element in rational choice theory. Some of the leading theorists of rational choice introduce their view by contrasting it with what they claim is the implicitly atheistic approach of social science tradition (Stark and Finke 2000). But the errors they make (e.g., their selective use of data when arguing that scientists remain religious) consistently serve to protect faith from secular criticism. Rodney Stark has even written against Darwinian evolution, which he accuses of being an antireligious ideology with inadequate scientific support (Stark 2004). Rational choice theory too easily shades into apologetics defending religious faith.

Rational choice theory does not appear to be an unqualified success. Nevertheless, it illustrates what a sociological view of religion should look like. Its virtues—making substantive predictions and conceiving of religion as a normal human activity—have to be reproduced by any theory that tries to do better.

THE SECULARIZATION THESIS

Rational choice theory seeks to supplant the secularization thesis, which observes that the public role of religion has been declining in modern Western societies and claims that this decline is due to social changes that accompany modernization.

Western Europe is the clearest example of secularization. Citizens of countries such as Britain, France, and the Netherlands no longer go to church often, and they admit little influence of organized religion into their everyday lives. The prestige and power of the churches has diminished; the state and the economy operate with no reference to religious convictions. Fine church buildings converted to warehouses or conference centers is not a surprising sight in Western Europe.

Many social forces can diminish the public role of religion. In modern states and economies, societies have fragmented, and life has become divided into more specialized enterprises. We are no longer just peasants, priests, or soldiers; we all occupy a bewildering range of occupations and social roles. So modern societies have a harder time sustaining a unifying religious vision. At the same time, modern life has less of a local focus, and larger groupings such as the nation have become more important than the close-knit communities that have always harbored the most intense religiosity. Large societies are also culturally more diverse. Awareness of this diversity and the need to do business with people of different religions fosters relativism. If people come to think that all religions have their own perfectly valid truths, that there is no unique way to salvation, their commitment to their own sects will weaken. Furthermore, the organization of modern societies embodies a bureaucratic style of rationality concerned with efficiency. In a world of bureaucracy and technology, the supernatural becomes more distant, less plausible (Bruce 1996).

Sociologists who associate modernity with secularization do not expect that people will always demand religion. There still might be a strong psychological basis for religiosity, but specific belief systems have to be learned, reinforced, and transmitted to the next generation. Religious traditions do not recreate themselves spontaneously. Religion is reproduced through collective social acts, not the whims of market forces. And modernization often interferes with the social mechanisms reinforcing and reproducing organized religion. A collection of individual religious consumers might continue to entertain supernatural beliefs, but these beliefs will not be as socially potent as those of an old-fashioned religious community.

So, according to the strongest form of the secularization thesis, the bulk of the people in a modern society will begin to lose interest in religion. As secularization proceeds, the supernatural becomes practically irrelevant to peoples' lives, hanging on as a kind of undemanding, therapeutic but socially inconsequential spirituality such as that found in liberal churches and the New Age. Individuals in a secular society do not turn into rationalist nonbelievers; in particular, the state of science has no direct effect on their level of religious participation. Instead, modern people become indifferent to religion. They might retain some weak supernatural beliefs—in Europe, for example, well over half of those surveyed still express at least a vague belief in a soul—but religion ceases to have consequences for worldly action (Bruce 2002).

Secularization is admittedly a limited perspective on religion, applying only to the modern, industrialized West. Even there, secularization is not supposed to proceed smoothly. Human societies are complicated; revivals of traditional religions, new religious movements, and interest in the individualist spirituality of the New Age will all happen alongside the decay of mainstream religion. But defenders of the secularization thesis expect that in the long run, the social significance of religion will continue to diminish. After each cycle of revival and backsliding, old-time religion will become weaker. New religious options will remain small, individualist movements that rarely attract intense commitment; they will not replace the void left by the shrinking churches. And although religion can find a new lease on life by taking on a role in defending a culture—as in Catholic Poland against communism or the ethnic immigrant enclaves in North America—that too will not turn the secular tide in the long run.

Broadly speaking, secularization seems a correct description of what has happened in Western Europe. There, the decay of organized religion appears irreversible. The strongest challenge to the secularization thesis is found in the United States. A much higher proportion of Americans believe in God than do Europeans, and they believe in a much more traditional God. More Americans participate in religious life. Even U.S. politics is strongly influenced by religion.

American religion is not entirely a picture of health. While conservative religiosity has remained prominent, it has done so at the price of becoming more individualistic, therapeutic, and consumer oriented. Vast numbers of Americans live essentially secular lives, even when they affirm vague beliefs about God, the Bible, heaven, or hell. Younger people today appear to be slightly less interested in organized religion than older generations (Duncan 2004). Still, Americans retain a moral consensus that religion is indispensable. A strong evangelical subculture enjoys significant political influence. In many parts of the United States, social life continues to revolve around church activities. So while religion in the United States shows some slight signs of losing ground, it is hard to see the same severe decline as in Europe. According to the secularization thesis, religion should decay when the social mechanisms that help reproduce a religious way of life do not function well. But this leaves open the possibility that religions can develop new social mechanisms to propagate belief and commitment. The continuing vitality of American religion suggests that the United States has been secularizing very slowly, that U.S.

churches have successfully developed new social mechanisms to keep themselves indispensable, or that the United States might be an exception that calls the secularization thesis into question.

Comparing secularization and rational choice theories reveals considerable controversy in the sociology of religion. Human societies are enormously complicated; no theory can claim to capture more than overall social patterns. Religious change will never be smooth. Nevertheless, sociology can answer some questions about religion. Social mechanisms, whether they are market forces or ways communities propagate their faith, shape the religious landscape. What particular religious expression flourishes or withers, what seems plausible or outlandish, all depend on whether religions can muster the social forces to reproduce themselves. In the United States, religious communities have had considerable success in creating a parallel culture with Christian music and media and supporting sectarian schools where students can be taught creationism and isolated from relativist influences. The religious Right has even gained government support for religious organizations through faith-based initiatives. These are exactly the sort of things that are necessary for old-time religion not to succumb to a secular consumer culture.

Nonbelievers generally hope that secularization accurately describes the trend in their societies. But however such controversies will be resolved, the theories in play in sociology have next to nothing to say about the reality of the supernatural. Furthermore, it also is clear that sociology alone cannot explain religion. Questions such as why most people believe in supernatural agents, why they make costly displays of belief, and why such beliefs are usually so socially central, would remain even if scientists were to map out the social mechanisms of religion to a much better degree than they do today.

RELIGION AND GROUP SELECTION

If there is more to religion than social forces, perhaps the deeper explanations lie in human nature. There is a common perception that human nature boils down to genetics, and so research indicating a weak genetic basis for religiosity has attracted popular attention, being presented as the discovery of a "God gene" (Hamer 2004).

A genetic component to religiosity does not mean there is a switch in the DNA that turns supernatural belief on or off. Religion should be rooted in the genes, since all human capabilities are enabled by our genome. The connections between genes coding for proteins and complex cultural behaviors are, however, indirect. Some people are musical, but others are tone deaf regardless of any training. Genetics has a lot to do with the cluster of talents that go into musical ability or its lack. Nevertheless, there is no gene specifically for musical ability. Similarly, it is plausible that many religious people are genetically predisposed toward more easily inferring invisible agents responsible for ambiguous stimuli. Many nonbelievers may be less able to perceive purposes behind everything. None of this implies that there are genes with an explicitly religious function.

The more interesting questions about religion and human nature have little to do with the genes themselves. Any genetic component to religiosity will naturally be interpreted by religious people as due to divine design: the gods allow some people to perceive divine realities. Nonbelievers see this research differently, as confirmation that religion has natural causes. The scientific question, though, is why human genes should enable religious behavior. Such questions are the business of evolutionary biology.

Evolutionary explanations of behavioral traits are similar to explanations of anatomical features. Traits that confer a reproductive advantage on individuals are more likely to be reproduced. Finding out what advantage religion secures for individual believers, however, is not a straightforward task. Some of the reasons are:

- Religion is a strikingly *collective* behavior. It is practiced by groups; human groups are often identified by their religion. Explaining collective behavior as arising from individual advantage is never straightforward. If a behavior is good for the group, an individual might enjoy the benefits without shouldering its share of the cost. It is often to the advantage of an individual to cheat, or ride free while letting others do the work. How, then, can collective behavior evolve?

- Religious behavior is *costly* and hard to fake. Believers have to devote significant resources to religious rituals and demonstrations of faith. Saving up for and going on a pilgrimage to Mecca is no small undertaking. All those resources going into religion are diverted away from behaviors that directly serve individual reproductive interests. Why does the believer go on a pilgrimage, rather than spending the money to help his sons and daughters get ahead in life?

- To outsiders, religious traditions often look strange, as they include what look like blatantly false representations of external reality. For example,

Figure 6.1 From an outsider's point of view, the rituals of a religious faith often seem odd, and the doctrines behind them hardly intelligible—for example, eating the flesh of a god. American Maryknoll Priests giving the Eucharist during Palm Sunday at Buhangija mission, Shinyanga, Tanzania. (Sean Sprague/The Image Works)

Roman Catholics and the Orthodox eat a wafer magically transformed into the flesh of their God (see Figure 6.1). At face value, believing in falsehoods is unlikely to confer any individual advantage. Religious beliefs are typically protected from easy empirical refutation, but the excuse-making process itself is a drain on resources. Religious concepts appear to be different than everyday beliefs that do genuine explanatory work. Why then devote resources to maintaining them?

Along with answering such questions, an evolutionary picture of religion must account for the specific, culturally universal features of religions. For example, supernatural agents predominate in religions. Even when almost impersonal powers such as karma are involved, they are personally and morally significant; in any case, popular religion is always full of gods and lesser spirits. Why can only personal agents do the work of religion? For that matter, what distinguishes the gods from pop-culture superheroes? (Atran 2002)

Biologists explain behavior that benefits a group but is costly for the individual in terms of inclusive fitness. Evolution is about replicating genes, but multiple copies of genes already exist in a population. So self-sacrificing behavior can be advantageous from the point of view of a gene, as long as the sacrifice boosts the reproduction of other copies of the gene. Such copies are more common in close relatives, so it is especially no surprise to see behavior benefiting kin. Relatives look after one another. Another kind of advantage of self-sacrifice comes when the individual can expect future rewards for self or kin due to the sacrifice. A soldier is more willing to die if he knows his family will be taken care of. We take much care to maintain a good reputation, to be known as reliable people who will reciprocate favors when called on—our welfare depends on it.

Cheating is always a problem with social animals. We are vulnerable to con men and shiftless relatives. Still, there are counterstrategies available. So we devote resources to detecting and punishing cheaters, thus raising the cost of cheating. There is an open-ended evolutionary arms race between better strategies for cheating and countermeasures to avoid being exploited. None of these are new issues for evolutionary biologists.

To explain religion in terms of inclusive fitness, biologists need to see how religiosity collectively benefits a group—to identify a social function for religion. Indeed, the role of religion has often seemed obvious: it is a social glue, holding human groups together. Religions embody norms. In separating the sacred from the profane, religions enforce useful behaviors appropriate to a group's environment. By defining morality, they determine ways to treat neighbors and enemies. Scientists who put an emphasis on religion as social glue take morality, and how religious behavior marks out who belongs to a group and who is an outsider, to be the central features of religion.

The strongest version of the view of religion as social glue portrays religion as a group-level adaptation, brought about through group selection (Wilson 2002). Normally, biologists focus on individual organisms. But according to multilevel selection theory, it can be useful to shift emphasis to groups in certain circumstances. If a species lives in cohesive groups, and these groups compete with each other, natural selection can operate at the level of group traits. If, for example, human bands engage in frequent warfare, it is a group-level advantage to have individuals fiercely loyal to others in the group and hostile toward outsiders. Each individual can gain an advantage within the group by cheating. However, if their reproductive success strongly depends on the success of their group in intergroup competition, any in-group advantage of cheating will be wiped out by the disadvantage to the group as a whole. It may benefit a person to steal from a neighbor, but not if her group members become more distrustful of each other, cannot organize effectively, and are massacred in the next raid by a rival band.

From such a perspective, it is tempting to think of human societies as a kind of superorganism, with traits that are shaped by group selection. Religion, with its emphasis on morality, loyalty, identity, ritual demonstrations of group membership, and so on, will seem to be the device our species of ape evolved in order to keep groups together. Religion exists because it has a vital social function.

Currently, such a biological approach is more a research project than a full-fledged theory; it leaves too much unclear about how genetically coded human propensities are translated into the cultural features of religion, and how these features are shaped under group selection. And it is very difficult to demonstrate that intergroup competition is so important in human evolution that group selection has a significant effect. Still, group selection has a plausible connection to current ideas about the coevolution of genes and culture (Richerson and Boyd 2005). It remains attractive to think of human groups as being somewhat similar to organisms and that religion should have a social function. And the group-selection approach is based squarely on the well-developed biological answer to the question of how collective behavior can be organized when all that matters in evolution is reproduction of individual genes.

Even so, the proposal that religion is a group-level adaptation has many difficulties. Among anthropologists, the idea that supernatural belief has a specific social function has long been out of favor, for good reasons. Even the notion that religion is a social glue is open to criticism: religion is often a strong contributor to social strife and fragmentation. Though evolutionary biologists are used to looking for functions and adaptive traits, this approach need not be as successful when dealing with culture.

Another problem is that for a biological account to take hold, social norms and religious beliefs would need to reliably manifest the genetic information that is ultimately what is subject to selection. With anatomical traits and even many behavioral patterns, such a connection is straightforward: genes have a direct role in constructing the body and wiring the overall structure of the brain. When it comes to social norms, however, it is questionable whether there is any such stability of expression or transmission.

An even more worrying problem is how biological accounts ignore the human mind when trying to explain religion. Biologists have assumed that the mind and its cognitive structure are just an intermediate link in the chain between gene and culture; the details of human minds could be overlooked if the ultimate cause of religion was its adaptive value. But, overlooking so much leaves major questions unanswered. Biological accounts say next to nothing about why supernatural agents predominate in religious belief—they have to say that faith in supernatural powers just happens to be a social glue for our species. But if the "why supernatural agents" question goes unanswered, biological accounts of religion must at least be severely incomplete.

There is no reason to doubt that the human capacity for religion is rooted in evolutionary biology. But humans seem to be predisposed to believe in powerful supernatural agents, and religions organize human interactions with such supernatural realities. The concept of group selection does not illuminate the details of why this is so.

A VIRUS OF THE MIND

Scientists looking for an evolutionary account of religion have an alternative to emphasizing biology. There is also the proposal that in the realm of culture, there is another replicator alongside genes: memes. Ideas themselves can be reproduced, with varying degrees of success, in human brains and perhaps in other information-carrying media as well.

Religion has naturally attracted much interest among thinkers who have tried to develop the idea of memes. In particular, memetics proposes an intriguing answer to the question of who benefits from religion. Religious behavior is puzzling because it so often seems costly to the individual believer. But from a meme's-eye point of view, there is no puzzle. The beneficiaries of religion are religious memes themselves, regardless of any disadvantage to the believer—the host of the memes.

According to memeticists, some memes replicate well because they benefit the host, such as knowledge of which mushrooms are safe to eat. In contrast, memeticists have tended to portray religious memes as more akin to viruses: information that exploits the host to propagate itself, regardless of the consequences for the host. For example, successful religions usually command believers to ensure that their descendants grow up in the faith. The most widespread religions do not limit their concerns to the orthodoxy of a chosen people, but enjoin their followers to convert the world. Individual believers gain some benefit from living in an environment where most share similar moral perceptions. Nevertheless, the resources devoted to evangelism are disproportional to such modest benefits, not to mention the cost associated with missionary activity, crusades, and jihads against the infidel. From a meme's-eye view, however, the missionary imperative is a very useful addition to a memeplex that is out to reproduce as widely as possible.

Religious memes employ other reproductive strategies aside from evangelism. Most religions do not encourage criticism of their central beliefs. They present faith, even blind obedience to religious authority, as a great virtue. Furthermore, they make it very difficult to test religious fact claims. It is hard to directly check on whether a virgin gave birth to a god long ago. Whether there is an afterlife of rewards or punishments is impossible to investigate with everyday tools. So religious beliefs protect themselves against mechanisms of criticism and correction to which other claims about the world are subject (Blackmore 1999, chapter 15).

If all this is correct, religions are memeplexes devoted to protecting their integrity and reproducing themselves throughout as many brains as possible. Some liken religion to a virus of the mind, typically infecting its hosts during childhood. To survive, a child must uncritically accept the words of her elders; it is too costly for each generation to rediscover which mushrooms are poisonous. This trusting behavior also leaves a window of opportunity for religions and other superstitions to infect the mind.

Describing religion as a virus is hardly the sort of thing the faithful would do. The memetics of religion has been developed by nonbelievers, and it appeals mostly to nonbelievers. The very idea of a meme was first proposed by Richard Dawkins (1989), a biologist well known for his outspoken atheism. The Internet Infidels, a major Web site serving nonbelievers, has a slogan, "Culture jamming theistic memes since 1995."

Nonbelievers can be tempted to treat religion as a virus, with its negative implications, while portraying nonbelief as the result of independent thought. However, as memeticists make clear, *all* ideas are supposed to be collections of memes, including the ideals of philosophical doubt or the scientific enterprise. A theory in physics is also subject to reproductive success or failure. If there is a difference, it is that the success of physics depends on its being responsive to reality tests and informed criticism, rather than its ability to protect itself by forbidding skeptical questions.

Memetics can take a more sophisticated role in nonbelief, beyond just labeling religion a virus. Most nonbelievers put a great emphasis on rationality. But if rationality is to run deeper than superficial means-ends reasoning as in rational choice theory, it must involve reflectively criticizing one's own purposes. Our personal interests are shaped by the reproductive interests of our genes and those memes lodged in our brains. Yet, on reflection, we might not want to act just as an appendage of our genes: we might favor other interests than just procreating and raising children. Similarly, we may prefer to be careful about which memes drive us, avoiding those that are too much like viruses. Critical rationality itself is a memeplex, however, so the rational ideal is not one of freedom from the replicators that make us but of favoring some over others (Stanovich 2004).

The difficulty with memetics is that its basis is still very speculative. Those who speculate about memes often leave them as ill-defined behaviors or information copied through people imitating one another. But that is not how minds work; imitated behaviors do not replicate with the kind of fidelity that allows Darwinian evolution to take hold. Communicated ideas do not trigger imitation but inferences common to our sorts of brain. If memes are going to be of any use to science, they must begin to be identified with brain structures, and their replication must be understood in terms of how ideas are generated in other minds through inference and evocation. In other words, memetics must be anchored in cognitive science. There are promising developments in this direction (Aunger 2002), so it is possible that the memetic approach has captured some solid insights into the nature of religion.

Making the connection to cognitive science is also critical for answering some questions that present meme-based approaches ignore. Memeticists have a mistaken tendency to treat the human mind as a blank slate that comfortably hosts just about any meme. The mind, however, has a preexisting structure—it is not just a featureless living space for competing memes. So memetics at present has no good answer to a question like why supernatural-agent concepts dominate in religion. Religions may have properties that help them reproduce well, regardless of what happens to their hosts. But why should, for example, the demand for blind faith most effectively combine with supernatural agents rather than with pink elephants?

THE RELIGIOUS MIND

Belief in the supernatural—in religious or paranormal realities—exists in all human societies. Religious ideas vary widely, from ancestral spirits to virgin births. Still, among extraordinary beliefs, not anything goes. No one believes that internal organs rearrange themselves within people while they sleep, and it seems obvious that such a belief is not a good candidate for religion (Boyer 2001). To find out what is a workable religious belief, and to be more precise about just what the term supernatural means, it seems reasonable to look to common features of human minds.

Evolutionary psychologists describe human minds as being built on a collection of more or less automatically activated modules. Due to our common genetic heritage, humans share an intuitive folk physics, a folk biology, a folk psychology, and other modules that allow us to make quick, usually successful judgments in a complicated environment. We classify objects we perceive in preset categories such as person, living thing, natural object, or tool. Each of these categories comes with a set of default expectations. Once we identify an object as an animal, we can immediately infer that it should reproduce within its own species, that it needs food, that it can die, and so on (see Figure 6.2). We do not need to be told such things: the expectations are automatic. If the object is a tool, we expect it to be manufactured and to serve a specific function. If it is a person, we activate the inferences of folk psychology.

Religious concepts violate the intuitive expectations of the category in which they belong. A spirit, for example, is a person, and most of the inferences we intuitively make about persons we also make about spirits. For example, they can communicate with us, their actions are understandable in terms of their purposes, and so forth. But we also expect persons to have bodies, and spirits violate this expectation. Supernatural agents appeal to common sense: religious concepts are easy to learn, and they evoke a host of inferences appropriate to the category of persons. Like commonsense beliefs, religious ideas occur to most people naturally, without need for much reflection. Yet supernatural beliefs are also odd, since they violate some basic expectations.

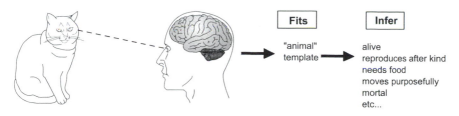

Figure 6.2 Humans automatically classify perceptions in preset categories such as "animal," and again automatically activate a set of inferences about what properties an animal can be expected to have. (Ricochet Productions; after Boyer 2001)

The stories told in human cultures are full of supernatural concepts—along with talking swords, walking statues, and other violations of intuitive object categories. Cognitive scientists have found that concepts incorporating a single violation are optimal for memory and oral transmission. The oddness grabs attention, while enough default inferences can be made to avoid having to remember too many details. It appears that the typical human brain easily remembers and transmits stories about category-violating beings. The next question, then, is what makes supernatural persons objects of worship—what makes us construct religions around them instead of just folktales?

Agents are a particularly important part of the environment for humans, who depend on social interactions for practically everything in life. It is critical for us to be able to detect agents. We do this imperfectly; after all, people behave in complicated ways, and we often have to figure out what people are up to indirectly, using incomplete information. Our built-in, intuitive agent-detection mechanisms can be dull, failing to pick out some of the purposeful activities of persons around us, or the mechanisms can be oversensitive, so we see faces in the clouds and a purpose behind every coincidence (see Figure 6.3). But since we are so dependent on the social environment, failing to detect real agents behind ambiguous stimuli is much more costly than being oversensitive and attributing a purpose to complicated but impersonal events. Humans, then, tend to be oversensitive: we are predisposed to see agents and purpose everywhere (Guthrie 1993). So category-violating agents such as spirits will not just grab attention and memory; they will also very often be validated by experience. Similarly, detecting predator-prey relationships is important for humans, and our intuitive detection mechanisms are easily triggered by uncertain, anxiety-producing events. Folktales and religious beliefs

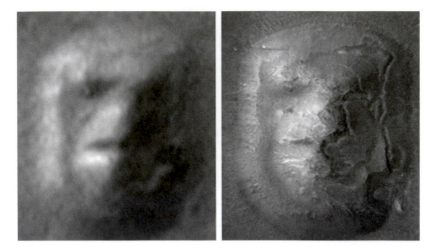

Figure 6.3 Our agent-detection mechanisms tend to be oversensitive. For example, we easily identify faces in random features, such as Jesus' face on a burnt tortilla. A recent example is the "face on Mars" in the left photo. In the 1976 Viking images a feature resembling a face attracted attention at the fringes of science, and some enthusiasts claimed it was really a face, a sign of alien intelligence. The 2001 Mars Global Surveyor image on the right looks much less like a face, but Mars Face believers insist that with suitable image enhancement and analysis it can still shown to be a face. (Images courtesy of Mark Carlotto and the *New Frontiers in Science* online journal)

alike are full of monsters, and worship of predatory, malevolent gods is as common as that of more beneficial spirits. The concept of supernatural agency is built on innate mental capacities concerning agent and predator detection (Atran 2002).

The role of the supernatural in our lives is solidified when such concepts are pressed into useful social work. We can expect to have social interactions with agents. In a social context, a particularly important type of supernatural agent would be one with perfect social knowledge. We all have some knowledge about our neighbors and spend a lot of effort gossiping and trying to find out more about other people's sexual behavior, whether they cheat on social obligations, what they keep secret. If a supernatural being is thought to have full access to important social information, interactions with such an agent should have very practical social consequences (Boyer 2001).

Religious rituals are a common way of organizing interaction with supernatural agents. Rituals are also usually costly and hard to fake—

although they may seem wasteful from the outside, they are very useful for demonstrating the believers' commitment. Rituals broadcast social information, signaling willingness to cooperate. Without a genuine belief in gods who will punish noncooperators, it is difficult to participate in the rituals and religious life of a community. So religion can help solve the problem all social groups face: to prevent cheating. Costly, hard to fake, publicly demonstrated commitment to gods who know everyone's secrets and will somehow punish cheaters is a good way to ensure cooperation.

Theories of religion based on cognitive science are continually debated and developed. Some aspects of a cognitive account command a consensus, such as how supernatural concepts should be understood as easily remembered and transmitted violations of innate categories. Others, such as how religion harnesses the emotional structure of human minds, and especially how much weight the fear of death carries in making religion a universal feature of human societies, are much less certain. Still, scientists today have a better claim than ever to an understanding of how religion is rooted in the structure of the human mind. This does not mean cognitive science can explain everything about religion. The details of how category-violating concepts came to figure in legitimating morality, of how societies developed priesthoods and orthodoxies, are historical matters. The large-scale organized religions of today cannot be understood apart from the social mechanisms by which they reproduce their doctrines. Still, social exchanges with supernatural agents, rituals, and the way the supernatural gets entangled with existential anxieties seem pretty basic, and cognitive scientists have made a good start on understanding such things.

In the cognitive picture, religion is not a simple thing that evolved, either through group selection of genes or cultural selection of memes. Religion involves a variety of cognitive systems with separate evolutionary histories, and none of these systems were selected primarily due to the social functions religion has come to serve. In other words, religion is an evolutionary byproduct. In our species, religious beliefs are almost unavoidable due to how our brains work—we usually believe in supernatural realities as unreflectively as we take more ordinary commonsense ideas about the world to be true. And as with any other evolutionary byproduct, religion can be co-opted to do useful work—in the case of religion, easing cooperation within a group. Nevertheless, religious beliefs are not directly *for* anything. Saying that religion is a social glue, that it is a prescientific way of explaining nature in terms of persons and purposes, or that religion eases existential anxieties may all be more or less true, but such observations do not explain religion. There is, from a cognitive science perspective, no single cause for religion, only a rough convergence of some features of human minds that make supernatural realities compelling for most of us.

EXPLAINING THEOLOGY AND NONBELIEF

Scientific explanations bring religion down to earth. If, for example, religion is a byproduct of evolved features of the human brain, then it becomes less plausible that religious beliefs arise from actual encounters with realities transcending the natural world. Still, sophisticated believers have some counterarguments. After all, science is also enabled by evolved features of the brain, and it is no less trustworthy for being so. Evolution may have endowed us with a reliable inclination toward the supernatural.

Moreover, explanations of religion are supposed to be very general, encompassing beliefs in witchcraft and ancestral spirits as well as the more abstract concepts of the world religions. It is tempting to say that scientists have so far sketched theories of superstitions and primitive religions, and that what they say does not apply to the indescribable, "wholly other" God of advanced philosophical theologies. Sophisticated people believe in rational theologies, not immortal monks or talking donkeys. Religion in the modern world is less about magic and more about morality.

There might be something to such objections. However, even in modern times, religion remains practical rather than abstract, magical rather than theological. The overwhelming majority of ordinary devout people see nothing to be embarrassed about in praying for healing, going on pilgrimages to tombs of saints, and taking miracle tales literally. God is supposed to be an all-powerful invisible person, not just the ultimate moral ideal or an almost meaningless "Ground of Being" that liberal theologians speak of. The attempt to make a coherent system out of unreflectively held supernatural beliefs is a preoccupation of intellectuals. They create theology, which takes shape at the intersection of religion and philosophy. Popular religion, however, does not depend on theology. So if current theories of religion do not fully account for philosophical theology, this is not an important limitation.

Even so, a cognitive approach to religion casts some light on theology as well. Much of theology is based on texts, and depends on logical argumentation rather than intuitively made inferences, but in its resistance to criticism and to reality checks, theology is not much different than unreflective religion.

Religious beliefs are represented as true—misfortune really is due to witchcraft, our soul does go to heaven or hell when we die. If we are able to think about beliefs as being true, however, then we can also conceive of their being false. Creatures with a brain sophisticated enough to deal with representations of truth and falsity can easily engage in deception, which is socially dangerous. Religious commitment should be hard to fake. Furthermore, although minimally category-violating concepts stick in the mind, they do not generate a coherent set of inferences—it is hard to give them a fixed meaning. If too many people understood that being protected by the spirits means they can walk off cliffs and be supported magically, there would be trouble. Believers must think religious concepts are true, but also limit how they act on their beliefs.

Successful religious concepts avoid such problems. One solution is vagueness or untestability. The violations of intuitive categories inherent in any supernatural concept mean they cannot generate a consistent set of inferences; it is hard to pin them down in order to test them (Atran 2002). In fact, direct testing becomes near impossible when outcomes like life after death are promised. Another solution is to validate religious claims in a context of absolute authority (Pyysiäinen 2004). A good Muslim is bound to equate It says so in the Quran with It is true; a critical scholar who deviates from this authoritative framework is automatically suspect. Conservative Protestants think the Bible is authoritative and self-interpreting; they insist that any criticism that does not make that presupposition misses the point. In either case, religion is protected from criticism.

Theology, no less part of the philosophical tradition than freethinking, aims to be critical and reflective. However, theology is also much concerned with apologetics, with protecting belief from doubt. To accomplish this task, theologians regularly try to insulate religious claims from criticism. They seek a context of absolute authority; when watered down, this becomes the liberal insistence that religion is sui generis, understandable only from the inside. They interpret super-natural claims in a metaphysical sense, and then treat metaphysics as an area invulnerable to reality tests of any sort. They argue that science is always compatible with religion, since science has nothing to say about the ultimate metaphysical truths revealed by religion. Although all this turns God into a cosmic Santa Claus, such defensive strategies are very often compelling to believers, largely because they exploit features already present in unreflective religiosity.

If explanations of religion illuminate much, from basic supernatu-ral concepts to aspects of rational theology, they should also have something to say about nonbelief. After all, nonbelief is also enabled by the structure of the human mind. And since cognitive science-based explanations of religion portray supernatural belief as coming natu-rally to human minds, nonbelief appears as an anomaly, an exception. Early attempts at formulating a psychology of religion were apt to take belief as the oddity that needed explaining. In speculations such as those based on Sigmund Freud's notion that religion was a mental aberration and God a father figure (Freud 1928), nonbelief appeared as the healthy position, a normal state that might be obscured by a re-ligious culture. But now it seems that nonbelief is the exception in-stead. So it becomes interesting to ask what causes some people to reject the supernatural.

A unified explanation for nonbelief might not exist, however. Evo-lutionary explanations focus on supernatural belief as a common fea-ture of human populations. But populations are made up of individuals with a range of variations. Just as some people are reli-gious overachievers, some have a more skeptical temperament. Reli-gious ideas will not be equally arresting to every brain. Village skeptics who dislike the way religious concepts violate intuitive cate-gories will likely always be with us; they need no special explanation.

Similarly, the fact that religion is so commonly entangled with morality means there will always be some opposition to religion as a legitimating ideology. Social groups require cooperation, but people still have conflicts of interest. Moral opposition to the local religion can be as gut level and unreflective as much belief, though it is more often expressed as religious deviation rather than as a complete re-jection of the supernatural. Nevertheless, moral opposition is impor-tant for nonbelief, at least as a starting point motivating people to question the authoritative context validating religious beliefs.

Reflective nonbelief—intellectual doubt—is the form most flatter-ing to the self-conceptions of most nonbelievers. Explanations of reli-gion do not tell us much about reflective nonbelief, other than that it is enabled by the structure of our brains like any other intellectual ca-pability. In fact, there has been very little solid research on the psy-chology and cognitive science of nonbelief in general. Until we know more, our understanding of religiosity itself will remain incomplete.

THE PROSPECTS FOR NONBELIEF

The social and behavioral sciences have been making good progress in explaining religion as a natural phenomenon. So intellectually, the social sciences mildly support nonbelief, in the sense that they extend a naturalistic picture of the world.

Nonbelief, however, is rarely just an intellectual position concern-ing the nature of the universe. Nonbelievers typically have moral and political objections to religion as well—they believe not just that su-pernatural claims are false, but that we would be better off if many fewer people seriously believed in their religious traditions. Many nonbelievers associate religion with irrationality and think people should aspire to do better. And just as churches commission socio-logical studies, nonbelievers can also look at social science with a hope of learning how to spread skepticism more effectively. In that case, the problem for nonbelievers is that none of the serious ideas in play in explaining religion give much reason to expect a majority to adopt a naturalist outlook rather than supernatural beliefs.

Rational choice theorists present belief as an economically rational option. They expect no long-term decline in the demand for religion, and to the extent that they pay any attention to nonbelief at all, it is as an unfortunate influence on sociology.

Nonbelief has better prospects according to the secularization the-sis, but not much better. Religion may be losing its social clout in West-ern Europe, but this is taking place largely due to cultural relativism, lack of interest in the older organized religions, and supernatural be-lief taking on diffuse, socially impotent forms. Science-minded reflec-tive nonbelievers are a small minority in Europe as elsewhere.

Biological explanations of religion tend to find religiosity inscribed in the human genome. If so, the consequences for nonbelief are un-clear. For example, a genetic tendency to nearsightedness might have once been a serious problem, but it is easily correctable in an environment where technology is available. An inclination to supernatural belief might also be easily swamped by culture and education. It is more likely, however, that religiosity is a much more resilient aspect of human nature.

Memetic views of religion are attractive for many nonbelievers; if religion is like a virus, maybe it can be cured. Unfortunately, the best approach to memetics, combined with an effort to use the rationality memeplex to help choose between memes competing for brain space, suggests that the sort of reflective rationality science-minded nonbelievers aspire to is difficult. More to the point, reflective rationality is costly. It takes a lot of training and background knowledge for a reasoned approach to work well, and even then, it is typically slow and ponderous. Approaches that deploy intuitive ways of thinking are quicker and demand fewer resources, so even if they lead to errors such as a belief in supernatural agents, they may end up enjoying an overall advantage in most everyday situations. So the prospects for rational nonbelief being reproductively successful beyond its current small niche are uncertain at best.

Another way for nonbelief to displace religion is in the form of a secular ideology, as Marxism once attempted. But secular ideologies and moral philosophies are always at a disadvantage when compared with religion, since they cannot harness the emotional intimacy of a relationship with a supernatural agent. Neither are they very good at addressing existential anxieties like the fear of death. Religion may be an evolutionary byproduct rather than an adaptation; nevertheless, human minds are strongly inclined toward the supernatural, and it is hard to see what else can replace a form of thought that does so much social and therapeutic work so easily (Atran 2002, chapter 10).

If religion is here to stay, so is nonbelief. Historically, philosophical doubt has been a viewpoint restricted to an educated elite; being reconciled to holding a minority viewpoint in a sea of religious humanity will not be new to the tradition of nonbelief. Even today, especially after the collapse of the secular messianic hopes of Marxism, many nonbelievers do not immediately hope for a triumph over religion. Instead, skeptics strive to maintain a social space for individuals to be able to live without religion. The more realistic hope for nonbelievers is a secular state and culture that does not require citizens to proclaim a belief to be considered morally trustworthy—an environment where religion has become a private matter. Secularization to that minimal extent might still be possible, even outside of Western Europe.

Scaling back the cultural ambitions of nonbelief is only reasonable, but it has its own dangers. All beliefs, all social forms have to reproduce themselves. And any set of ideas confined to an intellectual ghetto invites stagnation over time. Hence, especially among nonbelievers who find a moral virtue in resisting supernatural faiths, scientific explanations of religion will continue to evoke an ambivalent response. That such explanation is possible at all encourages nonbelief, but the best theories available deflate hopes of a nonreligious future for humanity.

NOTE

1. The surveys that provide evidence for this, however, are less adept at capturing sympathy for more diffuse, liberal supernatural beliefs. It is possible that natural scientists are more inclined toward the more concrete claims of both traditional religious doctrines and outright naturalistic atheism.

Chapter 7

~

Morality and Politics

MORE THAN THE FACTS

Debates over science and religion can range far beyond questions of fact. Nevertheless, skeptics and true believers both claim to have a good idea whether supernatural agents are responsible for the observed universe. Many religious thinkers try to enlist science to support the fact claims of their tradition. Nonbelievers do the opposite, and they also uphold science as a model of rationality, superior to faith-based thinking.

In the intellectual debate, nonbelievers have come to expect to have the upper hand. After all, naturalism has become pervasive in modern science. The established knowledge that appears in textbooks makes no reference to any spiritual reality. There are plenty of questions for which scientists have no settled answers, but even at the cutting edge of research, none of the proposals that have a good prospect of success support supernatural beliefs. Religion is alive and vigorous outside the scientific community, and so there are no end of attempts to get science to recognize spiritual facts. But without exception, such attempts—such as parapsychology or intelligent design—either fail outright or get relegated to a shadow existence at the fringes of mainstream science.

Science clearly challenges supernatural religion, and the intellectual response from conservative religious thinkers has not been adequate. Theological liberals have largely attempted to avoid the challenge by

downplaying the supernatural elements of religion and by arguing
that the scope of science and its naturalism are limited. But modern
science is ambitious. It tends to ignore limits, asking questions philosophers and theologians would like to have reserved to themselves.

Nonbelievers urge everyone to draw the obvious conclusion. Science is pervasively naturalistic, most likely because the supernatural
does not exist. The notion of realities beyond the natural world has
been thoroughly tested and found wanting. But to the endless frus-
tration of nonbelievers, few people are persuaded by such an argument.

Naturally, nonbelievers want to explain why their arguments have
such a limited effect. They often end up with a shrug, saying that religion has too strong a hold on believers. Such a response risks becoming a self-serving excuse, the way creationists explain away their
near-universal rejection by the scientific community by charging that
scientists are blinded by antireligious ideologies. Even so, it is true that
ideas such as naturalism are not judged solely according to how well
they explain our world. People have other interests in life than satisfying their curiosity, and few ideas are free of social consequences. So
the debate over science and nonbelief is shaped by political concerns
as well as intellectual content.

Consider the common claim that religion complements science by
revealing the deeper meaning of events. Science, the argument goes,
is blind to meaning, being only concerned with material causes and
natural explanations. A religious person can accept science without
reservation, enriching her understanding by taking the meaning dis-
cerned by a religious outlook and adding it on top of the explanation
provided by science. AIDS, for example, is certainly a viral disease,
correctly described by medical science. But the deeper significance of
AIDS, some religious people think, is that it is a divine judgment on
sinful people such as homosexuals and drug users. Science-minded
nonbelievers see no reason to accept such a hidden meaning. More
important, they approach the very idea of meaning differently. Rather
than a mysterious dimension beyond science, they consider meaning
to be a construction by material brains interacting in a natural world
(Dennett 1995, chapter 14). Nevertheless, whether the notion of religious meaning as an add-on to science will be compelling to large
numbers of people will not be determined by intellectual considerations alone. We can guess that religious conservatives will have limited enthusiasm since they do not want to entirely surrender the job

of judging fact claims to science. We can also say that liberals will favor an idea that promises to preserve the attractions of both science and religion. But to say more, we need to look at the social circumstances in which the claim of add-on meaning is being made.

Whether add-on religious interpretations are persuasive depends on the social authority of a religion. In a religiously diverse society, divine punishment will not be the only pronouncement on AIDS; some people of faith will take it as an opportunity to show selfless love for the suffering. While the faithful debate meaning, though, scientists can count on widespread agreement about basics. When the relevant technical experts agree that AIDS is caused by a virus, the scientific community quickly reaches a consensus.[1] Furthermore, most people are aware that scientists can achieve significant agreements. In contrast, disagreement about religious interpretations is common in modern societies. Different sects and religions will often disagree even about the most fundamental matters. So any add-on meaning provided by religion risks appealing only to narrow communities of faith. Because of general awareness of the ability of scientists to reach consensus, plus widespread respect for the power of technology, there will always be a demand for ideas promising to help science and religion complement one another. But the promised religious meaning either ends up being specific and sectarian, or inclusive and vague. Indeed, most liberal, inclusive pronouncements on deep meaning are glaringly superficial. In secular, pluralist societies, most everyone can ignore the add-on meaning proclaimed by modern religions without suffering practical consequences. The meaning religion offers becomes optional, enjoying only a limited social role (Bruce 2002).

Similar considerations affect the fate of any idea, including scientific and philosophical positions. Whether ideas catch on and how widely they are propagated are never due to intellectual merit alone. Scientific institutions are supposed to try and structure the rewards they provide in order to select ideas according to merit. Even so, they are never perfect. Everyday concerns such as the need to secure funding from government and corporate sources can hinder as well as help the search for truth. The academic and research-oriented world is not isolated from outside social and political influences. And outside academia, the standards are much lower. The media will always hype flawed studies of "the power of prayer" or promote UFO abduction tales, as long as such stories bring in audiences for advertisers.

Because of the pervasive naturalism of modern science, science and nonbelief stand intellectually close to one another. But nonbelief is not just an intellectual attitude. Nonbelief typically exists as a part of social movements with political ambitions, and science is no less a social enterprise entangled with politics. Whatever intellectual affinity they may have, science and nonbelief occupy different social positions and therefore can have conflicting interests. Hence, understanding the relationship of science and nonbelief requires more than an appreciation of evolution or the randomness in quantum physics. The political context influences everything, including the popularity of various views on science and nonbelief.

THE STATE OF SCIENCE

Scientists today should have little to complain about. By and large, science is well respected, and it commands significant public resources. There are hundreds of thousands of natural scientists active today, even when applied sciences such as medicine and engineering are excluded. In the United States, the National Science Foundation enjoys an annual budget of around $5.5 billion. When applied science is added to this picture, both the number of people engaged in scientific and technical activity and the resources devoted to their work increases manyfold. Technologically advanced nations typically devote between 0.5% and 1% of their government expenditures to research and development, and including private sector expenditures, the overall spending on research and development typically is over 2% of the gross domestic product (OECD 2004). Naturally, a large majority of such funds are devoted to weapons and health-related research and other applied science rather than basic natural science. Still, by any measure, science is a large-scale enterprise, able to sustain both small projects in thousands of modest laboratories and "big science" efforts such as putting telescopes in space (see Figure 7.1).

Workers in scientific fields accordingly enjoy status and income as professionals. Academics or researchers in basic science cannot expect to make as much money as medical doctors or stockbrokers, but they live comfortably. More important, they enjoy stimulating work and the respect of their neighbors. In other words, scientists do reasonably well in the present social order.

It is not, however, true that modern Western countries have scientific cultures. While most citizens consider science a worthwhile activity that usually benefits them, few can be considered scientifically

Figure 7.1   The Hubble Space Telescope is unberthed and lifted up into the sunlight during a servicing mission in 1997. Projects such as putting telescopes in space demand considerable public resources and therefore political support. (NASA Headquarters– Greatest Images of NASA, NASA-HQ-GRIN)

literate. Most of us are exposed to science in our formal education, and science courses are notorious for being among the least understood and most difficult hurdles the average student has to endure. Scientists and educators do a lot of hand-wringing about this state of affairs, usually concluding that our methods of doing science education are badly in need of repair. Unfortunately, getting a grasp of science beyond a superficial list of facts requires a lot of dedicated work. While it might be possible to have more people come to appreciate science from a distance, expecting a large proportion of the population to have even a minimal acquaintance with scientific thinking is probably unrealistic (Shamos 1995).

In that case, for most people in the foreseeable future, science will remain the arcane activity of a specialized group of experts. Scientists might enjoy respect but not informed trust—just the generic trust extended to those with relevant expertise. Few will appreciate science as a cultural achievement, and few will engage with science at any depth beyond watching the occasional nature program on television. Even fewer will be able to let modern science deeply influence their view of the world. In fact, with the extreme specialization of today's technical professions, even scientists feel little incentive to try and achieve an overall coherence in their understanding of the world. A chemist can be perfectly competent in polymer chemistry and still take homeopathic "medicine" when ill.

So while science is a well-established, adequately funded enterprise, its social support is thin. As an institution, science enjoys its position not because of widespread commitment to the intellectual aims of science, but because of the services science provides to powerful elites. Most important, science is entwined with technological innovation, which can be turned into military and commercial applications. Modern physics, for example, has become intimately connected to energy and weapons development. Medical and pharmacological research drives much of biology; chemistry today would be hard to imagine without funding from commercial interests.

In popular culture, science is identified with gadgets and confused with technology. Even groups distrustful of modern science as an intellectual enterprise, such as creationists, can feel very positive toward applied science. In fact, scientists who make a public case for continued support of science often encourage the confusion between science and the applications of science. When science enthusiasts argue that science has done great things, they almost always bring up medicine, not evolution or modern physics, though advanced medical knowledge is far less significant for our overall understanding of nature.

Since science receives support because of the practical benefits it promises, its political clout is limited. For example, in the United States, recent conservative administrations have pursued policies that have alienated many in the scientific community. Supporting creationism as an option in biology education, ignoring the scientific consensus on global warming, presenting false information about abortion and contraception, limiting stem cell research—such actions and more have led to deep concern among many scientists (Meyer 2004). However, compared with large corporations and religious conservatives, scientists and their friends are clearly a much less powerful constituency.

Science is further constrained by its advisory role in policy: scientific pronouncements are only valuable to the extent that they are impartial, untainted by politics. Appearing above politics while pursuing the political interests of science is not easy. Scientists need to convince powerful elites and the general public alike that scientific institutions must operate independently, free of political interference. Furthermore, basic scientific knowledge has to be a public good that is not restricted to any single corporation or military. Research and education in basic science demands public funds. So scientists, and particularly scientists in leadership positions, have to be concerned about the public image of science as well as the intellectual integrity of the scientific enterprise.

Appearing impartial and generally useful does not, however, prevent all conflicts of interest. Modern science continually encounters points of friction with religious conservatives and can also conflict with other political groups when issues of culture and morality come to the fore.

Religious conservatives often perceive the naturalism of modern science as a pernicious, secularizing influence on their societies. Their distrust of science flares up around morally charged claims. Religious conservatives are concerned that science, in its present form, is not just a neutral presenter of definite facts but a philosophically biased enterprise that wraps its facts in theories that undermine moral authority. Conservatives tend to validate the social order they prefer by portraying it as the natural, God-given order of things. Theories such as evolution undermine traditional notions of how a moral order is visible in creation. Worse, evolution, and naturalistic science in general, do not merely favor a rival moral vision, but undercut all attempts to read a moral vision out of the way the universe is structured. By portraying nature as devoid of any obvious underlying moral purpose, modern science gives credibility to moral relativism. Conservatives of all stripes see relativism as a social disaster that erodes the shared moral discipline necessary for a viable society. And religious conservatives rightly perceive that the relativism of modern culture is a more serious threat to old-fashioned religion than the nonbelief of a few intellectuals.

Business-class conservatives are less focused on morality, and have fewer deep-seated worries about science. Those who are ideologically committed to letting markets dominate all aspects of life dislike the notion of science as a public good, often arguing that science would be improved by privatization. Such conservatives are also often hostile to results from environmental science which suggest that corporate capitalism has led to a level of plunder of the planet that invites disaster.

Worries about science are not confined to the political right. Scientific naturalism also undermines more liberal and left-wing intuitions about how nature is infused with moral significance. In the world as described by science, people might not have any immediately obvious reason to go beyond self-interested behavior. Even Darwinian evolution derives creativity from individual competition and is easily described in terms of metaphors such as "selfish genes." Scientific materialism, to many, appears to justify the crass form of materialism where life is centered on selfish, material gain at the expense of all higher spirituality and self-sacrificing morality. Environmental problems, for example, might well be partly blamed on science, since science has stripped the sacredness from nature, leaving it vulnerable to be exploited at will. However, though such concerns lead to a certain cultural wariness about science, they rarely surface in the form of outright conflict.

A more explicitly left-wing critique of science points out that science today is part of the Establishment, charging science with being far too cozy with corporate capitalism in all its rapaciousness. Science claims a value-free impartiality, but its funding and research agendas are set by commercial and military interests. Many postmodern thinkers have gone even further, denying that science can be objective, and demanding that science should not be privileged over alternative "ways of knowing" that might produce more politically liberating pictures of the world (Parsons 2003). Following its immersion in identity politics, the cultural left holds forth the indigenous "science" of colonialized peoples, and even spiritual and occult belief systems, as examples of alternative, equally valid ways of knowing.

Even with such political points of friction, science in modern societies remains healthy. However, it seems science does have a cultural weakness, in that it is unclear how modern science fits into the moral life of communities. Friction with traditionalist constituencies is especially likely to challenge science, with the political ascendancy of religious conservatism in many parts of the world. In this environment, scientific institutions need to maintain a morally positive image, yet also act as a source of impartial advice. Being associated with nonbelief, in these conditions, is a liability for science.

THE STATE OF NONBELIEF

Today, people without supernatural beliefs find themselves in a politically uncertain position. Where there is religion, there will also be dissent from religion, and barring a collapse of intellectual culture, the philosophical tradition of doubt will continue to attract its share of adherents. The question about nonbelief has always been whether it can gain a broader social foothold. Socialist movements once held out the hope that ordinary people could be mobilized around a secular ideal of a just society, but those movements have collapsed or become demoralized. Only Western European social democracy has achieved a modest form of socialist ideals, and Western Europe is indeed the most secular part of the planet today. Among modern Western countries, the United States has taken an opposite direction, and is notable for the religious fervor of its population and its enthusiasm for devoutly conservative politics. In much of the rest of the world, South Asia, Africa, and Muslim countries especially, religion remains as important as ever in public life, and in many places is even increasing in significance (Norris and Inglehart 2004).

The politics of nonbelief starts everywhere from the realization that nonbelief is a minority position. Even in secular Europe, the decline of religious influence in social life means widespread indifference to religion, not active nonbelief. People who reject all supernatural claims, who have no room for souls and spirits in their view of the world, remain a minority. Outside Western Europe, though, nonbelievers also bear the burden of being a distrusted minority. An atheist in Europe is just someone who has another kind of personal belief about the nature of the universe. Elsewhere, including the United States, a nonbeliever is very often morally suspect. Respondents to a 1999 Gallup poll in the United States were asked whether they would vote for an otherwise qualified atheist candidate for office—at 49%, atheists were the least-accepted group; the next group, homosexuals, came in at 59%.

So outside Europe, active nonbelievers may have to worry about basic social acceptance, especially if they live outside enclaves where a secular culture predominates. In that case, we should expect that political action driven by nonbelief should be similar to other forms of identity politics, demanding equal respect in social and cultural life without having to compromise too deeply the unique characteristics of nonbelievers. Nonbelieving groups, however, face the extra difficulty that the nonreligious are notoriously difficult to organize.

Merely rejecting the supernatural is not enough of a common ground to make nonbelief a coherent identity. In the United States, nonbelievers are no more than 1 to 6% of the total population (Norris and Inglehart 2004), and few feel a need to come out of the closet or otherwise act as nonbelievers. Even the most prominent and long-standing among nonbelievers' organizations, such as the American Humanist Association, have a membership of only a few thousand. The constant disagreement among nonbelievers about labels also reflects their disunity. "Nonbeliever" or "unbeliever" sounds negative, in addition, it implies that the lack of religion is a deviation from the norm of religious belief. Alternatives include atheist, Atheist (insisting on capitalization), agnostic, nontheist, freethinker, humanist, secular humanist, skeptic, rationalist, bright, and so on, each emphasizing different aspects of dissent from religion and each provoking passionate debate among nonbelievers. Negative or not, "nonbeliever" remains the most accurate label.

American nonbelievers, then, make up a small constituency who, while often distrusted, are also generally comfortably off and usually do not feel much urgency to organize as a distinct interest group. Politically, nonbelievers are most commonly liberals and leftists (see Seidman and Murphy 2004), but there are significant numbers of nonbelieving right-wingers as well. Some conservatives argue that while intellectual nonbelief is all right and good for an elite, religion is indispensable as an instrument of moral discipline for the masses (Lewy 1996). Other conservative nonbelievers think of commercial markets as a model for all of society to emulate. Right-wing libertarians who consider all forms of government to be illegitimate restrictions on individual liberty include many individualist dissenters from religion among their numbers (see, for example, Smith 1991).

There is one political demand, however, that almost all active nonbelievers support: a secular state, or the separation of church and state. Nonbelievers naturally benefit from a secular state, and they usually think removing religious interference from government is an important milestone of human progress. Moreover, secular government goes hand in hand with a secularized culture where science, law, economic life, and so forth operate according to their own standards, independent of religious control. Such a broadly secular society favors the privatization of religious belief; individualist religions such as many forms of Protestantism and the New Age can flourish in such an environment. Nonbelievers therefore usually describe secularism as mere government neutrality toward religion, even as the best way to promote religious freedom. Liberal, individualist religious people usually agree. However, more communally oriented religions such as traditional Catholicism or Islam suffer a disadvantage in such a secular environment. In the United States, religious conservatives consider strict separation of church and state to unduly limit the freedom of faith communities to participate politically. So the height of the wall separating church and state is a continual matter for political debate, to the constant worry of nonbelievers.

Defending a secular state is further complicated by the fact that there are different versions of secularism. The English and American political tradition supports a secular government primarily as a way of keeping the peace in a religiously plural society. Citizens of English-speaking countries therefore are more receptive to pleas to support all religions against nonbelief, or to be sympathetic to Muslim immigrants demanding the right to live according to the norms of their religious community rather than one-size-fits-all secular laws. In contrast, the French tradition of secularism actively supports secularity as a basic republican value. In such a tradition, claims to universal human rights take precedence over group rights. For example, the French state more actively interferes with fundamentalist Muslim communities that treat women unequally.

Due to their strong interest in secularism, nonbelievers are immediately disturbed by religious interference with science and science education, such as demands to teach creationism in public school science classes. Conservative religious attacks on science are also attacks on secularism—on the integrity of a secular enterprise that should be able to operate independently of religious interests. And science is especially important to nonbelievers, as a symbol of the accomplishments of unfettered human reason, and because they perceive the results of modern science to support nonbelief (see Figure 7.2).

Still, for nonbelievers, science-related issues are rarely a major focus of activity. Even when nonbelievers criticize religion, arguments that science makes religion less credible are overshadowed by philosophical arguments, and even more, by commonsense objections to scriptures and religious morality. Politically and socially, the major hurdle for nonbelievers is to persuade the believing majority that those who deny the supernatural do not necessarily lack moral integrity.

Figure 7.2 Don Addis cartoon expressing concern about religious interference in scientific research, from *Free Inquiry* 24:4 (2004): 17. *Free Inquiry* is a leading secular humanist magazine.

A MORALITY OF REASON

Religious people, particularly conservatives, often associate lack of belief with moral confusion or even outright evil. Indeed, in religious communities, nonbelief is bound to raise questions. Religion is intimately tied up with moral perceptions and emotions, and it clearly helps in organizing a community with common moral views and a good deal of mutual trust. Breaking with religion sets the nonbeliever outside the moral community. Why should nonbelief then be respected? Can nonbelievers be trusted—can they assure their fellows that they recognize a higher value than self-interest? If they do not believe God is watching and will punish transgressors, what prevents nonbelievers from cheating when they think they can get away with it?

Religious thinkers elaborate on such worries. Naturalistic nonbelief, it seems, recognizes nothing transcendent, and so no ultimate purpose or meaning in life. If so, what, for a nonbeliever, provides the inner motive for being moral? Individual skeptics may very well be decent people, but if this is so, it is because they have internalized the religiously derived values of their culture. They seem to lack a deep, fundamental reason to be moral. Without some reality transcending nature, there can be no objective moral facts. Morality becomes a matter of subjective personal judgments, of culturally relative attitudes. How can an atheist say that Stalin, who committed mass murder in the name of an antireligious ideology, was evil? On what basis would a godless person condemn slavery as a universal evil, even in places where slavery finds widespread social acceptance? Nonbelievers in a religious environment can be good people, but a society of the godless would eventually degenerate into moral anarchy.

Nonbelievers reply that skepticism about religion involves a lot more than a bare rejection of faith. The philosophical tradition of doubt includes profound reflections on morality as well as the gods. There is an alternative, secular way: that of basing morals on reason (see Figure 7.3).

For millennia now, many philosophers have argued that moral principles could be established by reason, without the aid of revelation. Formulating abstract ethical principles and trying to show that reasonable people will accept such principles and do good to one another is a major philosophical enterprise. For example, as a guiding principle of ethics, utilitarians propose maximizing the happiness of as many moral agents as possible. Many others concentrate on principles of justice and fairness, such as John Rawls (1999) who says that reasonable, impartial people who had no idea of what social circumstances they were to be born into would choose to arrange their societies to benefit the least well-off as much as possible. No single principle commands the general agreement of philosophers. Still, ethics is one of the most prominent areas of philosophical inquiry, and in philosophical circles, moral questions are routinely discussed without any reference to theological commitments (Darwall, Gibbard, and Railton 1997).

Figure 7.3 Don Addis cartoon from *Free Inquiry* 24:4 (2004): 18. Nonbelievers are very concerned to argue that religion is not necessary for morality; many argue that non-religious morality is superior.

The search for abstract moral principles has many difficulties, how-ever. One is the question of motivation: why should a rational person choose to be moral in just the way a philosopher proposes? Someone inclined toward injustice might understand Rawls's argument, and yet point out that that no one can be totally abstracted from their ex-isting social circumstances. In a position of privilege, it just seems ra-tional to seek to maintain that privilege. Even if philosophers were to agree on what moral ideals such as justice were all about, what would compel a person to behave according to those moral ideals?

Another approach secular ethicists can take is to appeal not just to abstract principles but to instrumental rationality: enlightened self-interest. It might seem rational to cheat a neighbor, but only if there is a very good chance of getting away with it. In fact, a rational per-son will recognize that she is also vulnerable to cheating and try to arrange for a social order where cheaters are likely to be caught and punished (Baier 1995). Discouraging cheating is in everyone's interest because it allows everyone to more fully enjoy the benefits of cooper-ation. Basing ethics on enlightened self-interest solves the motivation problem: if cooperation is in someone's interest, they will not need an extra push to cooperate.

Still, appealing to rational self-interest is not enough to construct a secular ethic. After all, in a slave-owning society, the interests of slaves and owners conflict, but only the slaveholders have the power to act on their self-interest. It might be true that a society allowing slavery cannot do well when competing against free societies because of all the resources it devotes to the repression needed to maintain the slave order. If so, a rational outside observer might still prefer a free society that maximizes the benefits of cooperation. Unfortunately, such considerations cannot motivate people who always live inside particular societies with specific power imbalances. Moreover, in practice, there are lots of different ways to order human societies, and none seem to have a strong claim to be the best for all local conditions. So emphasizing rational self-interest can lead to a degree of moral relativism.

One way to reduce uncertainty would be to base morals on common human needs inherent in human nature. In fact, at the very origins of Western moral philosophy, as seen in the writings of Aristotle (1984), we find an argument for a universal morality based on nothing but readily available natural facts. Many secular philosophers continue to work in this tradition, making moral judgments that depend on a broad, commonsensical understanding of basic human needs. Furthermore, our historical experience can also be a guide in thinking about ethics. With basic moral perceptions honed by human evolution, and millennia of experience with moral reasoning—including ethical thought within various religious traditions—we have some idea what leads to human flourishing and what does not, regardless of uncertainties about the foundations of morals.

There are other options as well. So the philosophical debate continues, about virtue, duty, justice, and so on. Secular moral philosophy is not free of problems and open questions, so philosophers have no need to worry about unemployment any time soon. In practice, secular moral thought is eclectic: it draws on different philosophical approaches, unreflective moral feelings, and negotiations between conflicting interests. It muddles through. Nonbelievers can point out, however, that secular ethics is no worse than theological ethics in this regard. Religions may promise moral certainty, but even within the same overall tradition, moral disagreements are extremely common. Appealing to faith does not, in practice, solve anything. Reason is an indispensable tool for anyone who wants to resolve conflicts of interest without resorting to naked force. And so, in religiously plural societies, secular moral reasoning takes hold. Even fundamentalists present secular, utilitarian reasons to argue that the patriarchal family is ideal or that sexual expression should be disciplined. Pointing to a literally read Bible or Quran may suffice within the community of faith, but it carries little weight in the wider society.

So secularists are confident that morality is independent of theology. Nonbelievers can be as moral as anyone else, and philosophical reasoning can discover objective moral facts without bringing in the divine will. In fact, secularists often say that secular moral thought does better than religious doctrine in developing a humane perspective on moral questions. Modern advances in human freedom, such as enabling women to flourish outside the confines of the home, have been achieved against the bitter opposition of traditionally devout religious people.

What is notable in the tradition of secular moral philosophy, though, is how science has had only a minor role in the discussion. Nonbelievers typically want to enlist science in their cause, arguing that more knowledge will make us more capable of meeting human needs. A minority of nonbelievers are extremely enthusiastic about technology—whether as apostles of technocracy in the 1930s or extropians and transhumanists today, they expect technological development to provide salvation in this world. But even in such cases, science and technology remain tools serving prior moral purposes. They provide choices without illuminating the morality of the options.

This is not to say that moral philosophy is completely insulated from science. Some philosophers refer to current science to refine our notion of a moral agent. Some argue that if scientific naturalism is correct—so that people do not have the sort of magical free will our religious tradition imagines—our notions of moral responsibility and appropriate punishment should change significantly. Some refer to what science has to say about human nature to clarify what common "human needs" are; indeed, some of the most enthusiastic attempts to connect science and a universal morality come from nonbelievers who think Aristotle's approach was correct (Carrier 2005). But still, many philosophers remain dubious, thinking that science may tell us how the world is but says nothing about how it ought to be.

Separating scientific questions of fact from moral questions about values may be a good idea, but for the naturalistic variety of nonbelievers, there is a difficulty. Even if we cannot expect science to determine our moral choices for us, science should still help us explain human behavior. The good and harm people do, their gut-level feelings of good and evil, all should be capable in principle of being explained scientifically. So science has much to say about the nature of morality. Believers and nonbelievers alike, our commonsense way of thinking about good and evil is that we perceive objective moral facts about the world. We think such moral facts are, like the fact that the sky is blue, true independent of the interests of the parties involved. But if there is nothing more to the world than the sorts of objects and relationships described by natural science, where do moral facts come in? In a universe of particles bouncing around, moral facts seem nearly as out of place as a spiritual dimension of reality.

Intellectually, nonbelievers have drawn both on science and on the secular philosophy of ethics. For science-minded nonbelievers inclined toward complete naturalism, however, a tension between these two currents emerges. Naturalism stands against any transcendent reality. The philosophical tradition, however, has often elevated reason itself to a transcendent status, expecting to discover nonnatural facts such as moral truths through rational reflection. But if nothing is transcendent, if reason itself is but a cluster of abilities such as critical thinking, abstraction, and inference exhibited by embodied persons with material brains, how can moral facts be anything like those traditionally imagined?

THE SCIENCE OF MORALITY

To explain behavior, particularly moral behavior, today's science begins from an evolutionary perspective. Organisms act according to their interests; even a clam has reproductive interests that drive its limited behavior. In fact, reproduction is fundamental to explaining stable patterns of interests. According to biological evolution, we should commonly run into those behaviors that help reproduce their underlying genes.

Self-interested behavior is relatively easy to understand, so the question for evolutionary biologists has been accounting for altruistic, other-regarding behavior. The first step in moving beyond individual self-interest is to realize that it is genes that replicate, not individuals. Self-sacrifice on behalf of offspring or devoted help to close relatives makes good evolutionary sense, as more copies of one's own genes can thereby be transmitted to later generations. Since others share some of an organism's genes, individual interests are not isolated from one another. Genetic self-interest demands cooperation.

Another basic impetus for cooperation is the possibility of "I scratch your back, you scratch mine" arrangements between unrelated individuals. In social species with brains sophisticated enough to track social interactions, such reciprocal interactions for mutual benefit can get very complicated. Humans are good examples: people depend on their communities to such an extent that they have to rely on the cooperation of others whether or not they have had recent exchanges. So for humans, being known as a reliable reciprocator is a valuable asset, and being able to detect cheaters who might exploit cooperative attitudes is a vital skill. Much of our time is spent managing our own reputation as reliable cooperators and gossiping to figure out the reliability of others (Alexander 1987).

Though evolutionary biology uses plenty of metaphors that emphasize individual competition—survival of the fittest, the selfish gene—biology is also full of cooperation. Even an organism's genome is a set of genes that replicate together: they cooperate to such an extent that their whole reproductive success depends on one another.[2] Understanding competition and cooperation is very basic to life science. Indeed, theoretical biologists today have sophisticated tools to study how cooperation arises from reproductive interests, including game theory and complex computer simulations. Moreover, if other replicators such as memes are important alongside genes, the same sort of analysis applies. Cooperation is very often good strategy in pursuing individual interests, to the extent that in strongly social animals, it becomes hard to separate individual from group interests. From an evolutionary perspective, unselfish behavior is not a mystery (Sober and Wilson 1998).

An organism's interests shape its interactions with the world. Its perceptions serve reproductive interests by triggering appropriate actions. For example, a moving pattern of light or shadow might mean a predator to avoid or prey to catch. So perception does not separate fact and value; it is about attending to what is relevant to an organism's interests. It is no different with humans. Our acts typically follow gut-level perceptions of the right and wrong of a situation. We do not waste time and effort thinking hard about lists of moral principles; emotions of aversion or approval move us to rapid decisions that generally work well. We do have principles, but our moral systems are built on intuitive perceptions of good and evil. In the starkest examples such as confronting the torture of an innocent child, a normal person perceives evil as no less a fact than the color of the sky.

The human brain is not simply a collection of modules that work automatically and in parallel; we also have a slower, more deliberative way of thinking which can, sometimes, override quick intuitive reactions. So we also add moral reasoning on top as another layer conditioning our behavior. A few generations ago it was common for many Americans to express revulsion at the notion of marriage between black- and white-skinned people. Partly due to social and educational changes occasioned by reasoned criticism of racism, such gut-level revulsion has become more muted.

Humans are complex, and so any scientific account of moral behavior is bound to be complicated. There are still many unknowns. Nevertheless, many researchers are now confident that moral perceptions and behavior are rooted in biology and sculpted by social learning, and that we have come a long way toward understanding morality within the natural world. In that case, if we have a good scientific handle on the nature of morality, we can turn to some of the questions that have worried philosophers. In particular, how objective are the moral facts we perceive or reason our way toward?

An evolutionary picture of human behavior holds out some hope that we can discover real moral facts. After all, in doing science, we also begin with our evolved capacity to perceive facts about the world, such as the sky being blue. Modern science leads us to refine and even reject some of our ordinary perceptions, such as that the earth is flat and that the sun moves across the sky each day. Even so, it seems that we can start with commonsense intuitions about the world, subject them to criticism based on evidence, and develop extremely powerful, objective knowledge about the world. The same may be true with morality. We start with our intuitive perceptions of good and evil, subject them to criticism based on moral reasoning and experience, and hope to make progress. Our historical experience and developed moral philosophy may give us objective knowledge, such as that slavery is morally unacceptable.

Morality is a bit more complicated, however. After all, we can scientifically understand our subjective experience of blue as something that happens when our kind of eye and brain systems respond to light of a certain wavelength range. Light is among the constituents of the universe about which objective statements of fact can be made. In the

Figure 7.4 Different observers perceiving light with a blue wavelength activate their brains in similar ways, and judge the light to be blue in color. Their perceptions correspond to an objective reality. The same people may then observe a peace demonstration, but their brains may respond differently, judging it to be a good or a bad thing. Although the social reality is the same, secular thinkers inclined toward a degree of relativism allow that the observers' moral judgments may differ without anyone making an error of fact or reasoning. (Ricochet Productions)

world described by natural science, to what does a perception of good or evil refer? If unlike light, there is nothing "out there" that corresponds to good or evil, good and evil might remain only subjective reactions even after critical reflection (see Figure 7.4).

Those scientific naturalists who propose to understand morality objectively usually think that moral facts arise from facts about human needs and social realities, and that these are typically perceived as facts owing to the particular structure and evolution of our brains (May, Friedman, and Clark 1996). Evolutionary theory points out that some ways to advance a set of interests are objectively good or bad strategies. Cooperative behavior, a sense of fairness in social settings, and methods of preventing cheating are no surprise when they evolve, simply because these make up good strategies in a broad range of circumstances. Furthermore, humans have similar needs: food, shelter, respect, and so on. A good strategy to satisfy our needs is likely to be good for all of us. If we think of human morality as a way to negotiate the social world while pursuing human needs, then we can make sense of objective moral facts. Some ways of behaving are part of objectively good strategies in the social realm. And social reality is no less "out there" than wavelengths of light.

It is certainly true that our perceptions of good and evil are not completely arbitrary, since the social environment will select certain intuitions and discourage others. Not every perception serves an evolutionarily stable set of interests, and so, unsurprisingly, as a species humans share many moral intuitions. Even so, people also have deep, intractable moral disagreements. And thinking of morality as completely objective can be as misleading as it is to think of morality as nothing but arbitrary preferences.

Evolution can hone in on good strategies, but evolution also produces a lot of diversity. There is no single stable pattern of interests, no single species on which all life converges. Biological evolution results in complex ecologies with multitudes of species successful in their own niches, competing and cooperating with other species. In the social realm as well, we cannot expect that all people and groups of people will share a single pattern of interest, a single way of life. We should expect a moral ecology (Flanagan 2002). In other words, different people will often identify with different interests, have divergent moral perceptions, and even the best moral reasoning need not always lead them to a similar outlook on the good and evil of a situation. Unlike facts about blue light, facts about good strategies in the social realm can never be independent of the interests of various people. In the long run, not every strategy is good and not every pattern of interests is viable. So it is not true that without full-blown objective moral facts, anything goes. Nevertheless, the natural world does not include moral facts true for everyone regardless of their particular interests (Edis 2002, chapter 9).

Among naturalists, the status of moral facts is controversial—many philosophers continue to think that a robust sense of moral truth endures even in a fully natural world (Wielenberg 2005; Martin 2002). But even if they are correct, the nature of moral facts turns out to be different than imagined by most religions. Naturalistic morality is somewhat like cognitive neuroscience after it dispenses with the soul. Naturalists argue that even without souls, most of our ordinary understanding of human psychology remains in place—feelings,

thoughts, and so forth do not disappear when they are better explained within natural science. Nevertheless, many religious people
will think something very important will be missing from the world
if there are no immortal personal essences. Similarly, a naturalistic
view of moral truth cannot substitute for religious views of absolute
morality.

Furthermore, if the moral ecology view is more correct, and morality is neither subjective nor quite fully objective either, science-minded
people cannot wholly deflect accusations of moral relativism. Worse,
it appears science is not simply neutral, declining to endorse anyone's
strongly felt moral commitments. Modern science, if taken seriously,
complicates all moral certainties. What intuitively seem like solid
moral facts become, at best, strong moral agreements we can rely on,
and at worst, illusions hiding the real facts about conflicting interests.
Not everyone is greatly disturbed by such a view of morality; after all,
many of us are used to living with the uncertainties of modern times.
For many others with strong moral convictions, however, the hint of
moral relativism indicates a failure of science. If science cannot underwrite a completely objective morality, there are other ways of thinking, be it religion or traditional philosophy, that promise to do so.

Those naturalists inclined toward relativism can still point out that
a degree of relativism is not necessarily a bad thing (Harman 2000;
Garner 1994). After all, we are still able to engage in moral reasoning
to negotiate our social conflicts of interest. As with any viable social
institution, the moral enterprise itself inspires commitment in its participants and transmits this commitment to future generations. There
is always the hope that people working together can construct moral
agreements that are robust and reproducible.

Quite a few secular thinkers are sympathetic to such a view of
morality as a rational human construction rather than knowledge of
transcendent moral truths. In fact, it would seem that constructing
morality is just what the job of secular moral philosophy should be.
But morality is not just an intellectual exercise; it undergirds a way of
life. In particular, naturalistic nonbelief is rarely just a claim about the
nature of the universe—its proponents usually want to recommend
reflective rejection of the supernatural as an important part of a rational way of life. But if many ways of life are viable, it is possible that
nonbelievers are right about this being a godless universe, but wrong
if they think every rational person should reject the supernatural.

The problem for nonbelievers is that their ideal of critical rationality is costly to realize. It takes a lot of effort to understand modern science or to engage in secular moral philosophy. If another approach can serve human interests well enough most of the time, but at much less cost, that other way will enjoy a competitive advantage. Moreover, propagating and defending that other approach can itself become one of the interests people live by. Religion usually embodies just such an alternative to full-blown critical rationality. So as a way of life, reflective nonbelief might only work for a small minority of intellectuals and religious dissenters. For most people, religious approaches that mobilize intuitive notions of supernatural agents and exploit emotions that tie morality to such supernatural beliefs will be more advantageous. For nonbelievers, their way of life might work well; on critical reflection, they may find that their commitment to reason as they understand it is continually reinforced. But plenty of religious believers also find that their way of life works well for them, and reflective believers often admit that in the end, theirs is a self-reinforcing commitment of faith. In that case, all that can be said is that there are multiple viable, self-propagating ways of life available to us, and that each of them may well look perfectly consistent and satisfying to those who make such ways of life integral to their identity.

So bringing science to bear on the nature of morality leaves secular moral thinkers with a problem. It raises the possibility that morality is not arbitrary, but neither is it completely objective. As a result, many science-minded secular thinkers find that theirs is but one way of life in a complex moral ecology. This is a disturbing conclusion for those nonbelievers with ambitions to supplant religion with their version of reason.

SEPARATE SPHERES

Figuring out the nature of morality does not give much guidance for how either scientific institutions or nonbelieving social movements can present a positive moral image. Especially from a more relativist perspective, deep moral disagreements are no great surprise, and today's science provides few resources to suggest all rational people should be able to resolve their differences. In any case, achieving a secure social place either for science or for nonbelief is not a matter of solving an intellectual problem and then trying to have the political process implement the solution. Political interests shape the whole debate.

The most attractive position on science and religion today is that they have separate spheres. If science and religion have different areas of interest and competence, they need not conflict. For example, Stephen Jay Gould, the well-known biologist, argues that science and religion have "non-overlapping magisteria" (1999). Though not religious himself, Gould thought that religion was at its best when making moral pronouncements rather than trying to determine what the natural world looks like. Science, on the other hand, has nothing to say about what we ought to do. Provided that theologians do not insist on doctrines such as creationism, and evolutionary biologists do not jump to moral conclusions based on their work, science and religion should coexist with no difficulty.

A broad notion of separate spheres has long been the conventional wisdom, especially in more liberal circles. Science, it seems, is legitimately concerned with facts about the natural and social realm. Arguments about values, ultimate metaphysics, or the meaning of things are different; these debates are the territory of philosophers and theologians. Very often such a notion of separate spheres serves to protect liberal religion from science-based criticism. Still, even many nonbelievers can be content with leaving questions about religious matters to philosophy—they just think that the doubting tradition within philosophy makes a stronger case than the theological tradition. Moreover, though most active nonbelievers think people would be better off without religion, not a few among science-minded nonbelievers have no objection to religions if they are understood as moral traditions. Religions incorporate much social experience and practical wisdom; they certainly say more about morality than does science, which usually is either silent or may even hint at moral relativism.

Most theologians also like to avoid conflict with science. In fact, some radical theologians from the very liberal branches of Christianity and Judaism explicitly state that in the modern world, the supernatural is no longer believable by reflective, well-informed people (see, for example, Cupitt 2002). They take a "non-realist" view of God and similar religious concepts, taking them not to be concrete realities but symbols of our highest aspirations for a meaningful life and a humane morality. If radical theologians have any disagreement with nonbelievers, it is not about the nature of the universe but about the value of our religious traditions. Supernatural beliefs, radical theologians agree, are mistaken. Serious religion should be about meaning and morality in *this* world.

Most theologians are not so radical; they think considerations of meaning and morality hint at some ultimate metaphysical reality that transcends the natural world. Science cannot tell us about such things. Revelation, religious experience, and metaphysical philosophy are what count, and theologians usually believe theism can hold its own against the doubting tradition in philosophy.

Even so, it is notable how far the more liberal and sophisticated among theologians drift away from traditional doctrines. To avoid any explicit clash with science, they have to shy away from all claims about miracles, the creation and design of the universe and life, the existence of souls, and more. This is difficult, and very few can do it consistently across the board. And those who can manage it face the charge that they are trying to preserve the word "God" at the expense of anything real for which it could stand. Adding a theistic ultimate reality on top of the pervasively naturalistic universe of modern science is awkward at best—God becomes a cosmic Santa Claus. Moreover, since today's science attempts to explain how we value things or how we ascribe meaning to events, and does so in its usual bottom-up, naturalistic fashion, even reserving only morality and meaning to religion cannot work forever.

Intellectually, carving out separate spheres for science and religion is not easy. It is largely religion, not science, that comes under pressure, and religion has to retreat from all its distinctive fact claims. In a larger political context, however, separate spheres is a much more flexible idea, with some important advantages. The points of friction between science and liberal religion are remote from everyday life, and so are not the sorts of things that lead to institutional conflicts. Science and liberal religion can happily coexist and even join forces to resist conservative attacks on science and the modern world. Most professionals, secularized believers who think of religion as a private matter, and people who generally trust experts and educational institutions are attracted to a separate spheres view. After all, that view promises the best of both worlds.

To do its political job, however, separate spheres must remain a somewhat ill-defined notion. Ordinary liberal believers tend to be more conventional in their beliefs than theologians. Though they have little to do with the supernatural most of the time, they often still believe that the salvation history given in their scriptures is roughly accurate, that evolution proceeds under divine guidance, or that the big bang suggests a creator. In practice, the separate spheres viewpoint usually comes to mean that believers defer to science but add a religious gloss to scientific ideas. So keeping science and religion in separate spheres can be frustrating to strict nonbelievers who want to directly confront religion. Politically, however, a defanged liberal religiosity that seeks excuses for faith without interfering in the practice of science is ideal for scientific institutions.

A vague, somewhat superficial notion of separate spheres, then, is likely to remain the conventional wisdom about science and religion in modern societies. An unreflective separation is necessary to keep a loose coalition of secularists, liberal believers, and technical professionals together to resist fundamentalism and maintain a positive image for science. Since such groups tend to have the greater influence over secular educational institutions, they can make sure that education largely reflects this conventional wisdom.

The most significant challenge to the idea of separate spheres comes not from nonbelievers, but from rival ideas proposing that science and religion are not just compatible but mutually supportive. For example, New Age spirituality is ambivalent about science in its current form, but New Agers keep promising that a revolutionary "new science" will reunite the spheres of scientific and spiritual knowledge. Whether in parapsychology, alternative medicine, or holistic pseudophysics, New Age versions of science reaffirm spiritual realities. The New Age appeals to a similar population as the more liberal branches of traditional religions, and so its vision of a more spiritual science puts some strain on the modernist coalition supporting science's independence from religion.

The more potent challenge to the independence of science comes, however, from conservative religion. Many conservatives, such as creationists, think science should change in order to recognize the truth of their revelation. Since conservative religious movements enjoy much more support and political power than either the New Age or active nonbelief, public discussions of science and religion tend to be framed as a contest between liberal theology and a more traditionalist view that demands that religion should regulate all spheres of life.

Although maintaining a nonpolitical image is important for scientific institutions, they are closer to the liberal rather than the conservative side of today's struggles over the direction of modern societies. Hence, the vague version of separate spheres that helps keep a supportive coalition together remains an important idea, even if it is not always intellectually satisfying. By claiming to stick to their proper sphere, scientific institutions can defend their political interests, and maintain a morally more or less positive image without getting too deeply entangled in everyday political battles.

AN UNEASY ALLIANCE

Doing science does not require nonbelief. Plenty of scientists are faithful participants in their religious traditions. Though most believing scientists are content to let science and religion occupy separate spheres, some try to enlist science in the service of their faith. Even fringe movements that meet adamant resistance from mainstream science, such as the intelligent design movement, can count on support from at least a handful of scientists of undoubted competence. Scientists who are brilliant in the laboratory and also very active in a theologically conservative religion are not as common as in Victorian times, but even today, such scientists are not especially unusual.

Nonbelief does not require any deep scientific knowledge. There will always be people who doubt both the magical beliefs of popular piety and the mystery mongering of mainstream theologians. Some, especially if they come to think their local religion does more harm than good, will come to actively oppose faith. Science, however, need not play a large role in their nonbelief. And when it does, the "science" they refer to need not have much to do with actual science. On its Web site, the Raëlian Revolution bills itself as "the world's largest Atheist, non-profit UFO related organisation—over 60,000 members in 90 countries—working towards the first embassy to welcome people from space." Technically, it *is* an atheist organization, as they believe gods and angels are not supernatural beings but space aliens. Raëlians make much of science, proclaiming that science will replace conventional religion (Palmer 2004). But they also support intelligent design, saying that the correct naturalistic explanation of life is not that life evolved but that it was designed by extraterrestrials. Evidently, godless quasi-religious movements can compete with the best among fundamentalists in coming up with bizarre ideas.

Scientific and secularist organizations have different memberships, and they rarely work toward the same ends except when trying to resist religious pressure on science education and research. Scientific institutions are thoroughly secular, but only in the sense that fire departments are secular—they have goals and standards of performance that are independent of religious belief or nonbelief.

So either on an individual or an institutional level, science and nonbelief are clearly distinct from one another. In religiously conservative countries such as the United States, defenders of science and science education often have to reassure the public that science is no threat to religion, and so they repeatedly point out that many scientists are religious and that many religious leaders celebrate science. Science need not lead to nonbelief.

All this is correct; however, it is also misleading. Although science does not *demand* nonbelief, science does end up *supporting* the naturalistic variety of nonbelief.

Surveys are a poor substitute for argument, but it is clear that scientists are not a very devout population. It is true that many science-related professionals are religious, particularly if the more applied sciences are included. However, especially in the more basic sciences and among more elite scientists, indifference to religion and active denial of the supernatural predominates (Larson and Witham 1998). The pervasive naturalism within science can hardly be expected to have no effect on scientists' overall picture of the universe. In fact, scientists are often inclined toward a very strong form of naturalism, proposing that the world is made up entirely of physical objects and processes (Melnyk 2003). By and large, the world as pictured by modern science has no room for any ghosts or gods.

Active nonbelievers themselves are very positive toward modern science. A well-educated person who lives by a secular humanist ethic and who affirms science as a major accomplishment of human reason is much more representative of nonbelief than a member of a UFO cult, even if the cult replaces God with aliens. The tradition of the European Enlightenment is still the most intellectually significant strand in modern nonbelief. And the Enlightenment tradition has often been accused of being too uncritically positive about science. Whether such accusations have merit, it is certainly true that modern nonbelief will continue to draw on science both for its godless picture of the world and for its hopes to improve human lives.

Although nonbelievers affirm science, there are points of friction. Natural science strongly supports the nonbelievers' view that the world is not governed by personal forces. But the Enlightenment tradition has always had social ambitions aside from describing the world correctly. In particular, Enlightenment rationalists have hoped to establish a universal morality based on reason. Today's science does not underwrite any universal notion of morality; moreover, to some science-minded thinkers, evolutionary explanations of human behavior suggest that a degree of relativism about morality is correct. Enlightenment rationalists need not be too discouraged, since they can continue their work on constructing a universal, humane, rational morality and trying to persuade those people outside of the Enlightenment tradition. But there is still a practical problem: current scientific explanations of religion do not encourage any expectation that most people will become scientific naturalists. Supernatural beliefs appear to be too deeply ingrained, too socially useful. Nonbelievers have only Western Europe as an example of substantial secularization. And it is hard to predict whether the European social experiment will be a long-term success.

Even with some potential friction, however, nonbelievers can rightly claim intellectual support from modern science. Even relatively unsophisticated, popular arguments in favor of doubting religious beliefs regularly invoke science, asking if it any longer makes sense to explain anything by bringing up gods and demons. Without appeals to science, the case for nonbelief would be considerably weaker; philosophical and moral disagreements with traditional religion would be not be enough to persuade many to abandon spiritual beliefs altogether. Active nonbelief in the Enlightenment tradition is closely tied to science, and today's intellectual nonbelief continues to have no option but to embrace science.

From the point of view of scientific institutions, however, association with nonbelief is as often a nuisance as a benefit. Secular people who are indifferent toward or against religion are vital allies in defending Enlightenment ideals and political secularism. And a secular environment has so far been the best for the flourishing of science. On the other hand, too close a connection with nonbelief is also a political liability. Scientific institutions cannot afford to be perceived as undermining the values people care very deeply about.

So if there is an alliance between science and nonbelief, it is an uneasy one. Intellectually, science and nonbelief are not just compatible—they largely support each other. Politically, however, the relationship is somewhat one-sided. Nonbelievers need science much more than science needs nonbelievers. And maintaining the prestige of science, not nonbelief, is the interest both groups hold in common.

NOTES

1. Consensus need not be universal; there have been a marginal few AIDS dissidents with scientific credentials who dispute that AIDS is caused by HIV.

2. Genetic cooperation is not perfect; competition among genes also shapes the genome. Problems of enforcing cooperation and preventing free rides apply at the genetic as well as the social level; without largely overcoming such problems, large gene complexes reproducing together would not be possible.

Primary Sources

The selections here give examples of various science-based arguments that are used to support skepticism about religion and the supernatural. Only the first is historical, representing nineteenth-century materialism; the rest are contemporary arguments concerning physics and cosmology, evolution, materialist views of the mind, criticism of parapsychology, and a cognitive science-based approach to explaining religion.

— **1** —

Ludwig Büchner

Force and Matter, or, Principles of the Natural Order of the Universe. With a System of Morality Based Thereupon. **Translated from the 15th German edition; 4th English edition. London: Asher and Co., 1884, pp. 23–24, 210–212, 300–302, 481–483**

In his classic statement of nineteenth-century materialism, Büchner addresses many of the same areas of science as does today's literature on science and nonbelief. Many of Büchner's arguments continue to have currency today, even though the specific scientific disciplines he calls on have advanced vastly. Just as interesting, however, are those cases where nineteenth-century science has been superceded and Büchner is clearly wrong. For example, Büchner assumes a rigidly deterministic Newtonian physics and an eternal universe static in its large-scale features.

But, as we have said, no further proof is needed to demonstrate that matter is indestructible, and that it cannot therefore be created. How can that be created which cannot be annihilated? Matter must have been eternal, is eternal, and shall be eternal. "Matter is eternal; it changes only forms" (Rossmaessler).

The eternity of matter, or of substance, appears also from the following consideration: Science teaches us that an absolute vacuum cannot exist, while the infinity of space is set down by reason as axiomatic. Hence follows necessarily the conclusion that space must have been filled with matter, and that this must have existed from eternity. It follows, in addition, as was shewn in the preceding chapter, that the universe must be uncreated. A beginning and an end of the universe are as such inconceivable, and must be relegated to the limbo of spiritual or theological fancies.

This next selection concerns evolution. Büchner's argument still finds echoes today, though his understanding of evolution is far from our current picture.

This general law of variation, transformation and development, whatever may be the causes of the change in individual instances, being once laid down and recognized, we reach a firm standing-ground for the solution of the apparently almost insoluble question as to the "Whence?" of the organic world, and as to the natural causes of that which in the heading of this chapter we have termed "secular generation" or biogenesis, as the sequel of primeval generation. From the least promising beginning and from the simplest organic form-element, which the combination of inorganic materials evolved by spontaneous generation from the lowliest vegetable or animal cells, or even from a yet lower or yet more primal organic formation, that whole rich and multiform organic world which surrounds us at this day has developed itself progressively, in the course of endless periods of time, by the aid of natural phenomena. Whatever may have been the nature of the process of evolution as regards the details, however much may yet remain obscure and doubtful in regard to the exact manner in which the organic formation has taken place, this much at any rate we can aver with certainty: *that it has, and must have, happened without the interference of a supernatural power.* If at the present day this creation, while we survey the surrounding Nature, impresses us beyond measure, and if we cannot entirely repel the intellectual impression which points to the existence of a direct creative power, this feeling is in reality to be accounted for by the fact that we see the final results of natural forces that have worked through many millions of years spread out before us in *one* aggregate picture, and that, while we look only at the present, without remembering the past, it is difficult for us to imagine at first sight that Nature has evolved all this out of herself. And yet, there is no getting over it. Whatever may have happened in each individual instance, the general truth rests on irrefutable facts; there is the law of analogies, existing sometimes in the domain of embryology, sometimes in that of comparative anatomy; there are the prototypal organisms; there is the necessary connexion between the external conditions of the crust of the earth, and the origin and form of organic creatures; in fine, the gradual evolution of the higher organic forms out of the lower and lowest, keeping pace with changes in the development of the earth; and there is the paramount fact that the origin of organised beings was not an instantaneous process, but one extending throughout all geological periods. All these circumstances and conditions are indubitable truths, and are wholly incompatible with the idea of a personal and omnipotent creative power; such a power could not have contented itself such a slow, gradual and wearisome process of creation, nor could it have rendered the progress of its work dependent on the stages of the natural evolution of the earth.

The work of Nature, on the contrary, is the very antithesis of such a conception; it is wholly spontaneous, and consisting, as it does, partly of fortuitous and partly of necessary productions, it is infinitely slow, gradual, and climacterical. Therefore we cannot perceive in this work anything in the shape of a leap, pointing directly to a personal volition; form links itself on to form and transition onto transition.

Büchner takes a materialist position on the mind, though given the state of science in his time, his argument rested largely on criticisms of dualism.

As a last resource, spiritualism has hit upon the so-called *piano-theory*, according to which the mind stands in the same relation to the brain as a player does to his instrument. There is no escape through this loophole either. Who has ever heard of a piano that grows with its player, that lives with him, falls off and becomes ill with him, or by being out of tune makes him incapable of reflection, or continues to play confused melodies after the player has gone away, or which can only maintain its strength by constant change of material and a regular alternation of activity and rest? Such a piano would indeed be a remarkably strange thing, apart from many other difficulties which militate against that theory. To carry this monstrous comparison to its logical ends, we must admit the same or a similar proposition for every other organ of the body, and assign a nerve-soul to the nerves, a muscle-soul to the muscles, a liver-soul to the liver, etc., all rank absurdities, into which it is not worth while to enter any further. The word "mind" is nothing more than a collective word and a comprehensive expression for the whole of the activities of the brain and its several parts or organs, just as the word respiration or breathing is a collective word for the activity of the breathing organs, or the word digestion is a collective word for the activity of the digesting organ.

No doubt, in the case of the brain, that highest and fairest blossom of all terrestrial organisation, something more is meant than in the case of the organs of breathing or digestion; we are dealing with the highest achievement of material combinations, we might say with the intellectualisation of matter and with the life and destiny of all that is great and noble among man's achievements on earth. Everything comes *from* it, and everything proceeds *out* of it. It receives everything, and gives back everything. Who that has thrown but a single glance at the powers and tendencies of this most wonderful of all organs, of which unfortunately so many men scarcely know the proper use, can refuse to endorse what *Huschke* says: "In the brain lies the temple of the highest that is of interest to us. Yea, the destiny of the whole human race is indissolubly bound up in the 65 or 70 cubic inches of brain-mass, and the story of mankind is recorded therein, as in a vast book, full of hieroglyphic symbols."

Büchner also argues that morality can be explained within the natural world, and that nonbelieving morality is usually superior to that of religion.

So far from morality being incompatible with unbelief, it is like everything that man possesses, the outcome of a long series of acquirements handed down from generation to generation, and depends on definite natural and social conditions; it is therefore by no means the same throughout, or *semper eadem* as the Church of Rome calls itself, but by its very nature it is a product of growth and a thing that changes,—an expression of human knowledge, which proceeds and progresses with that knowledge itself. What we call "moral feeling" has its origin in the social instincts or habits which each human (or animal) society develops, and must develop within itself, if it is not to perish by its own incapacity. Morality, therefore, is evolved from sociability, or the faculty for living in a community, and it changes according as the particular ideas or necessities of any given society change. Thus, the nomadic savage thinks it is a very praiseworthy action to kill his father when effete with age, whereas in the eyes of the cultured European parricide is the most horrible of all crimes.

Now seeing that man is essentially a social being, and can, without society, either not exist at all or only be thought as a predatory animal, it becomes easy to understand that his living in social communion with others must have saddled him with duties of reciprocity which in course of time developed into definite moral axioms. The beginnings of this are to be found in family life, which in the sequel developed into tribal and national life. Morality is, therefore, much older than religion, the latter being only a requirement of the individual, while the former is a requirement of society and has its germ in the earliest beginnings of social co-existence. Thus, it stands to reason that morality cannot have originated in religion, but is entirely independent of it. It was not until a comparatively recent period of civilisation that the two became connected with each other, and by no means to the advantage of the former. For it may be averred without fear of contradiction that religion is injurious to morality, in so far as it assigns to it an aim based upon egotism and self-seeking, whereas pure morality finds, and ought to find, its reward in itself, so that it may subserve the objects of Society at large and be at the same time a blessing to the individual, as a member thereof. The original object of religious institutions was not, as has been admirably shewn by E. *Bournouf* in the history of creeds, to make moral or virtuous men, but merely to afford a simple corroboration of the metaphysical or supernatural theories invented by the ancestors. Many ages had elapsed before the different churches laid down definite rules of conduct for their members. In keeping with this, the ethnological researches of E. B. *Tylor* have shewn that the moral ideas of savages never and nowhere originate in religion, and that among them the touch existing between religion and morality is, as a rule, but very slight and only of secondary importance.

— **2** —

Steven Weinberg, "A Designer Universe?" *New York Review of Books*, October 21, 1999. Reprinted with the permission of *NYRB* and Steven Weinberg

Steven Weinberg shared the Nobel prize in physics in 1979 for his contributions to the development of electroweak theory. This article, based on a talk given in 1999 at the Conference on Cosmic Design of the American Association for the Advancement of Science, is a response to the growing popularity of claims that science and religion are coming together once again. Part of it is omitted due to length; in the part that has been omitted Weinberg argues that "on balance, the moral influence of religion has been awful."

I have been asked to comment on whether the universe shows signs of having been designed. I don't see how it's possible to talk about this without having at least some vague idea of what a designer would be like. Any possible universe could be explained as the work of some sort of designer. Even a universe that is completely chaotic, without any laws or regularities at all, could be supposed to have been designed by an idiot.

The question that seems to me to be worth answering, and perhaps not impossible to answer, is whether the universe shows signs of having been designed by a deity more or less like those of traditional monotheistic religions—not necessarily a figure from the ceiling of the Sistine Chapel, but at least some sort of personality, some intelligence, who created the universe and has some special concern with life, in particular with human life. I expect that this is not the idea of a designer held by many here. You may tell me that you are thinking of something much more abstract, some cosmic spirit of order and harmony, as Einstein did. You are certainly free to think that way, but then I don't know why you use words like "designer" or "God," except perhaps as a form of protective coloration.

It used to be obvious that the world was designed by some sort of intelligence. What else could account for fire and rain and lightning and earthquakes? Above all, the wonderful abilities of living things seemed to point to a creator who had a special interest in life. Today we understand most of these things in terms of physical forces acting under impersonal laws. We don't yet know the most fundamental laws, and we can't work out all the consequences of the laws we do know. The human mind remains extraordinarily difficult to understand, but so is the weather. We can't predict whether it will rain one month from today, but we do know the rules that govern the rain, even though we can't always calculate their consequences. I see nothing about the human mind any more than about the weather that stands out as beyond the hope of understanding as a consequence of impersonal laws acting over billions of years.

There do not seem to be any exceptions to this natural order, any miracles. I have the impression that these days most theologians are embarrassed by talk of miracles, but the great monotheistic faiths are founded on miracle stories—the burning bush, the empty tomb, an angel dictating the Koran to Mohammed—and some of these faiths teach that miracles continue at the present day. The evidence for all these miracles seems to me to be considerably weaker than the evidence for cold fusion, and I don't believe in cold fusion. Above all, today we understand that even human beings are the result of natural selection acting over millions of years of breeding and eating.

I'd guess that if we were to see the hand of the designer anywhere, it would be in the fundamental principles, the final laws of nature, the book of rules that govern all natural phenomena. We don't know the final laws yet, but as far as we have been able to see, they are utterly impersonal and quite without any special role for life. There is no life force. As Richard Feynman has said, when you look at the universe and understand its laws, "the theory that it is all arranged as a stage for God to watch man's struggle for good and evil seems inadequate."

True, when quantum mechanics was new, some physicists thought that it put humans back into the picture, because the principles of quantum mechanics tell us how to calculate the probabilities of various results that might be found by a human observer. But, starting with the work of Hugh Everett forty years ago, the tendency of physicists who think deeply about these things has been to reformulate quantum mechanics in an entirely objective way, with observers treated just like everything else. I don't know if this program has been completely successful yet, but I think it will be.

I have to admit that, even when physicists will have gone as far as they can go, when we have a final theory, we will not have a completely satisfying picture of the world, because we will still be left with the question "Why?" Why this theory, rather than some other theory? For example, why is the world described by quantum mechanics?

Quantum mechanics is the one part of our present physics that is likely to survive intact in any future theory, but there is nothing logically inevitable about quantum mechanics; I can imagine a universe governed by Newtonian mechanics instead. So there seems to be an irreducible mystery that science will not eliminate.

But religious theories of design have the same problem. Either you mean something definite by a God, a designer, or you don't. If you don't, then what are we talking about? If you do mean something definite by "God" or "design," if for instance you believe in a God who is jealous, or loving, or intelligent, or whimsical, then you still must confront the question "Why?" A religion may assert that the universe is governed by that sort of God, rather than some other sort of God, and it may offer evidence for this belief, but it cannot explain why this should be so.

In this respect, it seems to me that physics is in a better position to give us a partly satisfying explanation of the world than religion can ever be, because although physicists won't be able to explain why the laws of nature are what they are and not something completely different, at least we may be able to explain why they are not slightly different. For instance, no one has been able to think of a logically consistent alternative to quantum mechanics that is only slightly different. Once you start trying to make small changes in quantum mechanics, you get into theories with negative probabilities or other logical absurdities. When you combine quantum mechanics with relativity you increase its logical fragility. You find that unless you arrange the theory in just the right way you get nonsense, like effects preceding causes, or infinite probabilities. Religious theories, on the other hand, seem to be infinitely flexible, with nothing to prevent the invention of deities of any conceivable sort.

Now, it doesn't settle the matter for me to say that we cannot see the hand of a designer in what we know about the fundamental principles of science. It might be that, although these principles do not refer explicitly to life, much less human life, they are nevertheless craftily designed to bring it about.

Some physicists have argued that certain constants of nature have values that seem to have been mysteriously fine-tuned to just the values that allow for the possibility of life, in a way that could only be explained by the intervention of a designer with some special concern for life. I am not impressed with these supposed instances of fine-tuning. For instance, one of the most frequently quoted examples of fine-tuning has to do with a property of the nucleus of the carbon atom. The matter left over from the first few minutes of the universe was almost entirely hydrogen and helium, with virtually none of the heavier elements like carbon, nitrogen, and oxygen that seem to be necessary for life. The heavy elements that we find on Earth were built up hundreds of millions of years later in a first generation of stars, and then spewed out into the interstellar gas out of which our solar system eventually formed.

The first step in the sequence of nuclear reactions that created the heavy elements in early stars is usually the formation of a carbon nucleus out of three helium nuclei. There is a negligible chance of producing a carbon nucleus in its normal state (the state of lowest energy) in collisions of three helium nuclei, but it would be possible to produce appreciable amounts of carbon in stars if the carbon nucleus could exist in a radioactive state with an energy roughly 7 million electron volts (MeV) above the energy of the normal state, matching the energy of three helium nuclei, but (for reasons I'll come to presently) not more than 7.7 MeV above the normal state.

This radioactive state of a carbon nucleus could be easily formed in stars from three helium nuclei. After that, there would be no problem in producing ordinary carbon; the carbon nucleus in its radioactive state would spontaneously emit light and turn into carbon in its normal nonradioactive state, the state found on Earth. The critical point in producing carbon is the existence of a radioactive state that can be produced in collisions of three helium nuclei.

In fact, the carbon nucleus is known experimentally to have just such a radioactive state, with an energy 7.65 MeV above the normal state. At first sight this may seem like a pretty close call; the energy of this radioactive state of carbon misses being too high to allow the formation of carbon (and hence of us) by only 0.05 MeV, which is less than 1 percent of 7.65 MeV. It may appear that the constants of nature on which the properties of all nuclei depend have been carefully fine-tuned to make life possible.

Looked at more closely, the fine-tuning of the constants of nature here does not seem so fine. We have to consider the reason why the formation of carbon in stars requires the existence of a radioactive state of carbon with an energy not more than 7.7 MeV above the energy of the normal state. The reason is that the carbon nuclei in this state are actually formed in a two-step process: first, two helium nuclei combine to form the unstable nucleus of a beryllium isotope, beryllium 8, which occasionally, before it falls apart, captures another helium nucleus, forming a carbon nucleus in its radioactive state, which then decays into normal carbon. The total energy of the beryllium 8 nucleus and a helium nucleus at rest is 7.4 MeV above the energy of the normal state of the carbon nucleus; so if the energy of the radioactive state of carbon were more than 7.7 MeV it could only be formed in a collision of a helium nucleus and a beryllium 8 nucleus if the energy of motion of these two nuclei were at least 0.3 MeV—an energy which is extremely unlikely at the temperatures found in stars.

Thus the crucial thing that affects the production of carbon in stars is not the 7.65 MeV energy of the radioactive state of carbon above its normal state, but the 0.25 MeV energy of the radioactive state, an unstable composite of a beryllium 8 nucleus and a helium nucleus, above the energy of those nuclei at rest. This energy misses being too high for the production of carbon by a fractional amount of 0.05 MeV/0.25 MeV, or 20 percent, which is not such a close call after all.

This conclusion about the lessons to be learned from carbon synthesis is somewhat controversial. In any case, there is one constant whose value does seem remarkably well adjusted in our favor. It is the energy density of empty space, also known as the cosmological constant. It could have any value, but from first principles one would guess that this constant should be very large, and could be positive or negative. If large and positive, the cosmological constant would act as a repulsive force that increases with distance, a force that would prevent matter from clumping together in the early universe, the process that was the first step in forming galaxies and stars and planets and people. If large and negative the cosmological constant would act as an attractive force increasing with distance, a force that would almost immediately reverse the expansion of the universe and cause it to recollapse, leaving no time for the evolution of life. In fact, astronomical observations show that the cosmological constant is quite small, very much smaller than would have been guessed from first principles.

It is still too early to tell whether there is some fundamental principle that can explain why the cosmological constant must be this small. But even if there is no such principle, recent developments in cosmology offer the possibility of an explanation of why the measured values of the cosmological constant and other physical constants are favorable for the appearance of intelligent life. According to the "chaotic inflation" theories of André Linde and others, the expanding cloud of billions of galaxies that we call the Big Bang may be just one fragment of a much larger universe in which Big Bangs go off all the time, each one with different values for the fundamental constants.

In any such picture, in which the universe contains many parts with different values for what we call the constants of nature, there would be no difficulty in understanding why these constants take values favorable to intelligent life. There would be a vast number of Big Bangs in which the constants of nature take values unfavorable for life, and many fewer where life is possible. You don't have to invoke a benevolent designer to explain why we are in one of the parts of the universe where life is possible: in all the other parts of the universe there is no one to raise the question. If any theory of this general type turns out to be correct, then to conclude that the constants of nature have been fine-tuned by a benevolent designer would be like saying, "Isn't it wonderful that God put us here on Earth, where there's water and air and the surface gravity and temperature are so comfortable, rather than some horrid place, like Mercury or Pluto?" Where else in the solar system other than on Earth could we have evolved?

[. . .] With or without religion, good people can behave well and bad people can do evil; but for good people to do evil—that takes religion.

In an e-mail message from the American Association for the Advancement of Science I learned that the aim of this conference is to have a constructive dialogue between science and religion. I am all in favor of a dialogue between science and religion, but not a constructive dialogue. One of the great achievements of science has been, if not to make it impossible for intelligent people to be religious, then at least to make it possible for them not to be religious. We should not retreat from this accomplishment.

— 3 —

Richard Dawkins, "Obscurantism to the Rescue," *Quarterly Review of Biology* 72:4, 1997, pp. 397–399. Reprinted with the permission of the University of Chicago Press and Richard Dawkins

The biologist Richard Dawkins is possibly the best-known advocate of non-belief among scientists today—he is certainly among the most outspoken and least compromising when opposing religion. This article was a commentary on Pope John Paul II's message to the Pontifical Academy of Sciences in 1996, concerning evolution. Dawkins's preferred title was, "You Can't Have It Both Ways: Irreconcilable Differences."

A cowardly flabbiness of the intellect afflicts otherwise rational people confronted with long-established religions (though, significantly, not in the face of younger traditions such as Scientology or the Moonies). S. J. Gould, commenting on the pope's attitude to evolution, is representative of a dominant strain of conciliatory thought, among believers and nonbelievers alike:

Science and religion are not in conflict, for their teachings occupy distinctly different domains . . . I believe, with all my heart, in a respectful, even *loving* concordat [my emphasis].

Well, what are these two distinctly different domains, these "Nonoverlapping Magisteria" which should snuggle up together in a respectful and loving concordat? Gould again:

The net of science covers the empirical universe: what is it made of (fact) and why does it work this way (theory). The net of religion extends over questions of moral meaning and value.

Would that it were that tidy. In a moment I'll look at what the pope actually says about evolution, and then at other claims of his church, to see if they really are so neatly distinct from the domain of science. First though, a brief aside on the claim that religion has some special expertise to offer us on moral questions. This is often blithely accepted even by the nonreligious, presumably in the course of a civilized "bending over backwards" to concede the best point your opponent has to offer—however weak that best point may be.

The question, "What is right and what is wrong?" is a genuinely difficult question which science certainly cannot answer. Given a moral premise or a priori moral belief, the important and rigorous discipline of secular moral philosophy can pursue scientific or logical modes of reasoning to point up hidden implications of such beliefs, and hidden inconsistencies between them. But the absolute moral premises themselves must come from elsewhere, presumably from unargued conviction. Or, it might be hoped, from religion—meaning some combination of authority, revelation, tradition, and scripture.

Unfortunately, the hope that religion might provide a bedrock, from which our otherwise sand-based morals can be derived, is a forlorn one. In practice no civilized person uses scripture as ultimate authority for moral reasoning. Instead, we pick and choose the nice bits of scripture (like the Sermon on the Mount) and blithely ignore the nasty bits (like the obligation to stone adulteresses, execute apostates, and punish the grandchildren of offenders). The God of the Old Testament himself, with his pitilessly vengeful jealousy, his racism, sexism, and terrifying bloodlust, will not be adopted as a literal role model by anybody you or I would wish to know. Yes, *of course* it is unfair to judge the customs of an earlier era by the enlightened standards of our own. But that is precisely my *point!* Evidently, we have some alternative source of ultimate moral conviction which overrides scripture when it suits us.

That alternative source seems to be some kind of liberal consensus of decency and natural justice which changes over historical time, frequently under the influence of secular reformists. Admittedly, that doesn't sound like bedrock. But in practice we, including the religious among us, give it higher priority than scripture. In practice we more or less ignore scripture, quoting it when it supports our liberal consensus, quietly forgetting it when it doesn't. And, wherever that liberal consensus comes from, it is available to all of us, whether we are religious or not.

Similarly, great religious teachers like Jesus or Gautama Buddha may inspire us, by their good example, to adopt their personal moral convictions. But again we pick and choose among religious leaders, avoiding the bad examples of Jim Jones or Charles Manson, and we may choose good secular role models such as Jawaharlal Nehru or Nelson Mandela. Traditions too, however anciently followed, may be good or bad, and we use our secular judgment of decency and natural justice to decide which ones to follow, which to give up.

But that discussion of moral values was a digression. I now turn to my main topic of evolution, and whether the pope lives up to the ideal of keeping off the scientific grass. His Message on Evolution to the Pontifical Academy of Sciences begins with some casuistical double-talk designed to reconcile what John Paul is about to say with the previous, more equivocal pronouncements of Pius XII whose acceptance of evolution was comparatively grudging and reluctant. Then the pope comes to the harder task of reconciling scientific evidence with "revelation."

Revelation teaches us that [man] was created in the image and likeness of God . . . if the human body takes its origin from pre-existent living matter, the spiritual soul is immediately created by God. . . . Consequently, theories of evolution which, in accordance with the philosophies inspiring them, consider the mind as emerging from the forces of living matter, or as a mere epiphenomenon of this matter, are incompatible with the truth about man. . . . With man, then, we find ourselves in the presence of an ontological difference, an ontological leap, one could say.

To do the pope credit, at this point he recognizes the essential contradiction between the two positions he is attempting to reconcile:

However, does not the posing of such ontological discontinuity run counter to that physical continuity which seems to be the main thread of research into evolution in the field of physics and chemistry?

Never fear. As so often in the past, obscurantism comes to the rescue:

Consideration of the method used in the various branches of knowledge makes it possible to reconcile two points of view which would seem irreconcilable. The sciences of observation describe and measure the multiple manifestations of life with increasing precision and correlate them with the time line. The moment of transition to the spiritual cannot be the object of this kind of observation, which nevertheless can discover at the experimental level a series of very valuable signs indicating what is specific to the human being.

In plain language, there came a moment in the evolution of hominids when God intervened and injected a human soul into a previously animal lineage (When? A million years ago? Two million years ago? Between *Homo erectus* and *Homo sapiens*? Between "archaic" *Homo sapiens* and *H. sapiens sapiens*?) The sudden injection is necessary, of course, otherwise there would be no distinction upon which to base Catholic morality, which is speciesist to the core. You can kill adult animals for meat, but abortion and euthanasia are murder because *human* life is involved.

Catholicism's "net" is not limited to moral considerations, if only because Catholic morals have scientific implications. Catholic morality demands the presence of a great gulf between *Homo sapiens* and the rest of the animal kingdom. Such a gulf is fundamentally antievolutionary. The sudden injection of an immortal soul in the time line is an antievolutionary intrusion into the domain of science.

More generally it is completely unrealistic to claim, as Gould and many others do, that religion keeps itself away from science's turf, restricting itself to morals and values. A universe with a supernatural presence would be a fundamentally and qualitatively different kind of universe from one without. The difference is, inescapably, a scientific difference. Religions make existence claims, and this means scientific claims.

The same is true of many of the major doctrines of the Roman Catholic Church. The Virgin Birth, the bodily Assumption of the Blessed Virgin Mary, the Resurrection of Jesus, the survival of our own souls after death: these are all claims of a clearly scientific nature. Either Jesus had a corporeal father or he didn't. This is not a question of "values" or "morals," it is a question of sober fact. We may not have the evidence to answer it, but it is a scientific question, nevertheless. You may be sure that, if any evidence supporting the claim were discovered, the Vatican would not be reticent in promoting it.

Either Mary's body decayed when she died, or it was physically removed from this planet to Heaven. The official Roman Catholic doctrine of Assumption, promulgated as recently as 1950, implies that Heaven has a physical location and exists in the domain of physical reality—how else could the physical body of a woman go there? I am not, here, saying that the doctrine of the Assumption of the Virgin is necessarily false (although of course I think it is). I am simply rebutting the claim that it is outside the domain of science. On the contrary, the Assumption of the Virgin is transparently a scientific theory. So is the theory that our souls survive bodily death and so are all stories of angelic visitations, Marian manifestations, and miracles of all types.

There is something dishonestly self-serving in the tactic of claiming that all religious beliefs are outside the domain of science. On the one hand miracle stories and the promise of life after death are used to impress simple people, win converts, and swell congregations. It is precisely their scientific power that gives these stories their popular appeal. But at the same time it is considered below the belt to subject the same stories to the ordinary rigors of scientific criticism: these are religious matters and therefore outside the domain of science. But you cannot have it both ways. At least, religious theorists and apologists should not be allowed to get away with having it both ways. Unfortunately all too many of us, including nonreligious people, are unaccountably ready to let them get away with it.

I suppose it is gratifying to have the pope as an ally in the struggle against fundamentalist creationism. It is certainly amusing to see the rug pulled out from under the feet of Catholic creationists such as Michael Behe. Even so, given a choice between honest-to-goodness fundamentalism on the one hand, and the obscurantist, disingenuous doublethink of the Roman Catholic Church on the other, I know which I prefer.

— 4 —

Owen Flanagan,
The Problem of the Soul: Two Visions of Mind and How to Reconcile Them. **New York: Basic Books, 2002, pp. 3–8. Reprinted with the permission of the Perseus Book Group**

Owen Flanagan is a philosopher and cognitive neuroscientist. He defends a materialist view of the mind, which he also argues is close to certain Buddhist views. This selection is from a book where he explores the moral and religious implications of modern neuroscience.

DESOULING PERSONS

There is no consensus yet about the details of the scientific image of persons. But there is broad agreement about how we must construct this detailed picture. First, we will need to demythologize persons by rooting out certain unfounded ideas from the perennial philosophy. Letting go of the belief in souls is a minimal requirement. In fact, desouling is the primary operation of the scientific image. "First surgery," we might call it. There are no such things as souls, or nonphysical minds. If such things did exist, as perennial philosophy conceives them, science would be unable to explain persons. But there aren't, so it can. Second, we will need to think of persons as part of nature—as natural creatures completely obedient and responsive to natural law. The traditional religious view positions humans on the Great Chain of Being between animals on one side and angels and God on the other. This set of beliefs needs to be replaced. There are no angels, nor gods, and thus there is nothing—at least, no higher beings—for humans to be in-between. Humans don't possess *some* animal parts or instincts. We *are* animals. A complex and unusual animal, but at the end of the day, another animal.

It is no surprise that the images offered by perennial philosophy and science conflict, even though many of the details of the scientific image are not yet in place. Perennial philosophy teaches that we are spiritual beings and that everything turns on perfecting our spiritual nature. The scientific image is committed to desouling us. If there is room for human perfection within the constraints of the scientific image—and I think there is—it cannot be spiritual perfection as traditionally conceived, because we do not possess spiritual components.

As the third millennium opens, some are trying to diminish the conflict between the humanistic image and science by assigning them different domains and roles. No modern version of the Councils of Constantinople, in which various popes and emperors met to define church doctrine and condemn challenging heresies between the fourth and fourteenth centuries, has negotiated this wary standoff. The terms of this unspoken cease-fire call for a division of labor between the humanistic image as defined by perennial philosophy and science. The human sciences reveal our animal nature—something perennial philosophy has always acknowledged is there—but they have nothing to say about our full nature and place in the cosmos, conceived along the lines of the Great Chain of Being. From the point of view of those inspired by perennial philosophy (and this includes most scientists and philosophers, as well as ordinary intelligent folk), proponents of the scientific image, as I have described it, are playing out of their depth, behaving like the "know-it-all" adolescents many of us remember being—never lacking in confidence, often loud and brash, but seriously deficient in humility and wisdom. Predictably, the defenders of the scientific image in turn regard the defenders of the humanistic image as harmless but somewhat annoying old farts.

Since in fact the human sciences cannot deliver a picture remotely as detailed as that offered by perennial philosophy—the latter having had, among other things, a three millennium head start—defenders of the humanistic image can easily perceive science as akin to an annoying advertisement that makes incredible promises and is worth ignoring.

When I speak of the missing details of the scientific image, I don't just mean the small details, such as which neurotransmitter fixes memories in the brain, or the medium-size details such as whether the mind is best described as running one general-purpose computer program or numerous special-purpose ones, but the big details as well: nature versus nurture, for example, or whether we are sneaky egoists like our chimpanzee relatives or warm, cuddly, and largely peaceful like our bonobo relatives. All three levels of details are missing at present. Without them, the scientific image is more of a scheme for an image than a richly detailed picture or map. When it overreaches, it is seen in the eyes of detractors as just a "philosophy" in the pejorative sense—a mere theory or an advertisement for one.

Some might think that this ideological standoff is, under these circumstances, a sensible one. But I don't think so, for three reasons.

First, the scientific approach is now delivering some of the goods that it has promised. It becomes harder each day to dismiss the scientific discoveries that are in fact yielding new knowledge about human being. Martin Heidegger used the neutral word *dasein*—which means "being" or literally "there-being"—to indicate that our understanding of human being, of the nature of persons, and of what and how we are, is still uncertain. Heidegger was not one smitten by science, but in seeking a neutral word to speak of our way of being, he acknowledged that the scenario offered by perennial philosophy is deficient—at once sketchy and incomplete despite its long life, excessively flattering and thus inauthentic, and possibly logically incoherent in places. He was right. When we examine perennial philosophy with the standards of logic and evidence that it has developed and encourages us to use, we see that it fails to produce a robust and authentic picture of human being.

Second, the conflict between the humanistic image—as refined and endorsed by perennial philosophy—and the scientific image affects the lives of ordinary people, even if they are unaware of it. This conflict is not like many others between competing scientific or philosophical hypotheses, which can be resolved in the halls of academe and whose eventual resolution may not significantly affect how ordinary people think about themselves or the world. We live in the space of images and conflict between these images has consequences for how we live. Many people have felt Dostoyevsky's disquieting worry—"If there is no God, then everything is allowed." If the perennial philosophy were entirely secure, if it were firmly held and believed with assurance, this is not a thought one could even have. To be sure, this thought is more likely to arise from the news that there are Nazis among us than that there are neurons within us—more likely to arise, as it did for Dostoyevsky, as well as Nietzsche, Marx, and Freud, from noticing that disordered persons and politics abound than that science dispenses with God. But the fact is that the cumulative discoveries of the human sciences in the last century and a half, the combined forces of psychology, sociology, anthropology, primatology, evolutionary biology, genetics, and neuroscience, significantly affect the way ordinary people think and feel about themselves. And many have noticed that the human sciences appear to have better resources to explain how there could be Nazis among us than does perennial philosophy.

Third, the scientific image has developed methods to sort out most of the multifarious triggers and mechanisms that make us tick. Mind science, especially neuroscience, is fast maturing. Why does that make such a difference? Because there is no longer any place for the soul to hide. The mind, as conceived by perennial philosophy, has been toppled off the pedestal from which it supposedly performed its magic through mysterious capacities of free will to transform and manipulate—possibly to override—the combined forces of the natural and social environment, genes, and whatever else came its way.

We now live in the age of mind science. The first President Bush dubbed the 1990s "the decade of the brain" in expectation of dramatic research breakthroughs and medical applications. Prior to this growth in brain research the brain was largely *terra incognita*. I graduated from college in 1970 and considered going to graduate school to study the human mind. For me this meant studying the human brain. But in those days that pretty much meant observing the behavior of rats in mazes before and after removing or destroying various brain parts. I hate rats, but that aside, the whys and wherefores of rats confused about mazes hardly seemed like a promising approach to understanding the human mind. In 1970, behaviorism still reigned in many quarters, and behaviorists rightly referred to the mind/brain as a "black box."

Now everything is different. For the first time in history, the proponents of the scientific image can see into the black box. Cognitive science and cognitive neuroscience have reliable methods and tools for examining and identifying the way the mind/brain works. It is not as if we know in remotely complete detail how the mind/brain works. But we know this much: The mind/brain does its magic through the operation of neurons, with axons and dendrites that form synaptic connections, and via electrical and chemical processes that mediate attention, remembering, learning, seeing, smelling, walking, talking, love, affection, benevolence, and gratitude. René Descartes, a thinker as attuned as anyone could be in the seventeenth century to working out a picture of persons that is compatible with science, famously believed that we think with immaterial minds. He was wrong: The brain working in concert with the rest of the nervous system is our *res cogitans*—our thinking stuff. We are fully embodied creatures. Genes, culture, and history work through and with this extraordinarily complex tissue to make us who we are.

It is not as if we now understand human nature. However, what we can say with confidence is that we have a good sense of what needs to be explained—human nature and behavior—and of how to go about explaining it. There are many unknown forces in genes, mind, and culture that will affect the story we eventually tell about what it means to be a person, about why we think, feel, and behave as we do. But no scientifically minded person thinks we will need resources beyond those available to genetics, biology, psychology, neuroscience, anthropology, sociology, history, economics, political science, and naturalistic philosophy to understand the nature of persons. The beliefs in immaterial minds, and minor and major spirits, are in need of explanation. But spiritual forces will not do any explaining. Unless God and the angels are tampering with us, unless astral forces and extraterrestrials are messing with our brains and perceptions, science will one day be enough to provide a true picture of our *dasein*. Or—better, synthetic scientifically inspired philosophy will do so. This new philosophy will modify and, where necessary, displace and replace the perennial philosophy.

So the wary standoff between humanistic and scientific perspectives cannot be—in fact, is not being—maintained. Mind science is exploding, and in concert with the other human sciences it is overturning the traditional conception of mind. The mind is nothing like what perennial philosophy says it is. Something—the humanistic image as endorsed by perennial philosophy, it seems—has to give.

— 5 —

Susan Blackmore, "What Can the Paranormal Teach Us about Consciousness?" *Skeptical Inquirer* 25:2, 2001, pp. 23–25. Reprinted with the permission of the *Skeptical Inquirer* magazine, www.csicop.org

Susan Blackmore is a former parapsychologist who became convinced of the reality of psychic phenomena following an out-of-body experience, but who grew increasingly skeptical as her experiments found no paranormal phenomena but plenty of "wishful thinking, self-deception, experimental error, and even an occasional fraud." She no longer works on

the paranormal. In this article, she argues that parapsychology says nothing useful about consciousness; in the excerpt, she surveys the field and
explains why she is skeptical. It has also been edited to remove most of
her copious citations of the literature; these can be found in the original.
"Psi" is shorthand for psychic power.

I would love to be able to provide a fair and unbiased assessment of
the evidence for psi and decide whether it exists or not. But this is
simply impossible. Many people have tried and failed. In some of the
best debates in parapsychology the proponents and critics have ended
up simply agreeing to differ or failing to reach any agreement. The
only truly scientific position seems to be to remain on the fence, and
yet to do so makes progress difficult, if not impossible.

For this reason, if for no other, you have to jump to one side or other
of the fence—and preferably be prepared to jump back again if future
evidence proves you wrong. I have jumped onto the side of concluding that psi does not exist. My reasons derive from nearly thirty years
of working in, and observing, the field of parapsychology. During that
time various experimental paradigms have been claimed as providing
a repeatable demonstration of psi and several have been shown to be
false. For example, in the 1950s the London University mathematician
Samuel Soal claimed convincing evidence of telepathy with his special subject Basil Shackleton, with odds estimated at 10^{35} against the
effect being due to chance. These results convinced a whole generation of researchers and it took more than thirty years to show that Soal
had, in fact, cheated. Promising animal precognition experiments
were blighted by the discovery of fraud and the early remote viewing
experiments were found to be susceptible to subtle cues which could
have produced the positive results. As [Ray] Hyman puts it, "Historically, each new paradigm in parapsychology has appeared to its designers and contemporary critics as relatively flawless. Only
subsequently did previously unrecognized drawbacks come to light."

THE GANZFELD EXPERIMENTS

The most successful paradigm during that time, and the one I shall
concentrate on, has undoubtedly been the ganzfeld. Subjects in a
ganzfeld experiment lie comfortably, listening to white noise or
seashore sounds through headphones, and wear halved ping-pong
balls over their eyes, seeing nothing but a uniform white or pink field

(the ganzfeld). By reducing patterned sensory input, this procedure is thought to induce a psi-conducive state of consciousness. A sender in a distant room, meanwhile, views a picture or video clip. After half an hour or so the subject is shown four such pictures or videos and is asked to choose which was the target. It is claimed that they can do this far better than would be expected by chance.

The first ganzfeld experiment was published in 1974. Other researchers tried to replicate the findings, and there followed many years of argument and of improving techniques, culminating in the 1985 "Great Ganzfeld Debate" between Honorton (one of the originators of the method) and Hyman (a well-known critic). By this time several other researchers claimed positive results, often with quite large effect sizes. Both Hyman and Honorton carried out meta-analyses but came to opposite conclusions. Hyman argued that the results could all be due to methodological errors and multiple analyses, while Honorton claimed that the effect size did not depend on the number of flaws in the experiments and that the results were consistent, did not depend on any one experimenter, and revealed certain regular features of ESP. In a "joint communiqué" they detailed their points of agreement and disagreement and made recommendations for the conduct of future ganzfeld experiments.

The ganzfeld achieved scientific respectability in 1994 when Bem and Honorton published a report in the prestigious journal *Psychological Bulletin*, bringing the research to the notice of a far wider audience. They republished Honorton's earlier meta-analysis and reported impressive new results with a fully automated ganzfeld procedure—the Princeton autoganzfeld—claiming finally to have demonstrated a repeatable experiment.

Not long afterwards Wiseman, Smith, and Kornbrot suggested that acoustic leakage might have been possible in the original autoganzfeld. This hypothesis was difficult to assess after the fact because by then the laboratory at Princeton had been dismantled. However, Bierman carried out secondary analyses which suggested that sensory leakage could not account for the results. Since then further successes have been reported from a new ganzfeld laboratory in Gothenburg, Sweden, and at Edinburgh, where the security measures are very tight indeed. The debate continues.

How can one draw reliable and impartial conclusions in such circumstances? I do not believe one can. My own conclusion is based not just on reading these published papers but also on my personal experience over many years. I have carried out numerous experiments of many kinds and never found any convincing evidence for psi. I tried my first ganzfeld experiment in 1978, when the procedure was new. Failing to get results myself I went to visit Sargent's laboratory in Cambridge where some of the best ganzfeld results were then being obtained. Note that in Honorton's database nine of the twenty-eight experiments came from Sargent's lab. What I found there had a profound effect on my confidence in the whole field and in published claims of successful experiments.

QUESTIONS ABOUT THE GANZFELD RESEARCH

These experiments, which looked so beautifully designed in print, were in fact open to fraud or error in several ways, and indeed I detected several errors and failures to follow the protocol while I was there. I concluded that the published papers gave an unfair impression of the experiments and that the results could not be relied upon as evidence for psi. Eventually the experimenters and I all published our different views of the affair. The main experimenter left the field altogether.

I would not refer to this depressing incident again but for one fact. The Cambridge data are all there in the Bem and Honorton review but unacknowledged. Out of twenty-eight studies included, nine came from the Cambridge lab, more than any other single laboratory, and they had the second highest effect size after Honorton's own studies. Bem and Honorton do point out that one of the laboratories contributed nine of the studies but they do not say which one. Not a word of doubt is expressed, no references to my investigation are given, and no casual reader could guess there was such controversy over a third of the studies in the database.

Of course the new autoganzfeld results appear even better. Perhaps errors from the past do not matter if there really is a repeatable experiment. The problem is that my personal experience conflicts with the successes I read about in the literature and I cannot ignore either side. I cannot ignore other people's work because science is a collective enterprise and publication is the main way of sharing our findings. On the other hand I cannot ignore my own findings—there would be no point in doing science, or investigating other people's work, if I did. The only honest reaction to the claims of psi in the ganzfeld is for me to say "I don't know but I doubt it."

Similar problems occur in all areas of parapsychology. The CIA recently released details of more than twenty years of research into remote viewing and a new debate erupted over these results. Whenever strong claims are made critics from both inside and outside of parapsychology get to work—as they should—but rarely is a final answer forthcoming.

These are some of the reasons why I cannot give a definitive and unbiased answer to my question "Are there any paranormal phenomena?" I can only give a personal and biased answer—that is, "probably not."

But what if I am wrong and psi does really exist? What would this tell us about consciousness? A common view seems to be something like this: If ESP exists it proves that mental phenomena are independent of space and time, and that information can get "directly into consciousness" without the need for sensory transduction or perceptual processing. If PK (psychokinesis) exists it proves that mind can reach out beyond the brain to affect things *directly* at a distance, i.e., that consciousness has a power of its own.

I suspect that it is a desire for this "power of consciousness" that fuels much enthusiasm for the paranormal. Parapsychologists have often been accused of wanting to prove the existence of the soul, and convincingly denied it. I suggest instead that parapsychologists want to prove the power of consciousness. In philosopher Dan Dennett's terms they are looking for "skyhooks" rather than "cranes." They want to find that consciousness can do things all by itself, without dependence on a complicated, physical, and highly evolved brain.

I have two reasons for doubting that they will succeed. First, parapsychologists must demonstrate that psi has something to do with consciousness and they have not yet done this. Second, there are theoretical reasons why I believe the attempt is doomed.

[. . .] This is why I doubt that evidence for psi, even if it is valid, will help us to understand consciousness.

— 6 —

Pascal Boyer,
"Why Is Religion Natural?" *Skeptical Inquirer* 28:2, 2004, pp. 27–31. Reprinted with the permission of the *Skeptical Inquirer* magazine, www.csicop.org

Pascal Boyer is a cognitive anthropologist. This excerpt is from an article where he presents his particular cognitive science-based view of religion. While he sees religion as a natural phenomenon—gods and spirits are not real—he expects religious beliefs to remain "stable and salient in human cultures."

People do not generally strive to believe six impossible things before breakfast, as does the White Queen in Lewis Carroll's *Through the Looking-Glass*. Religious claims are irrefutable, but so are all sorts of other farfetched notions that we never find in religion. Take for instance the claim that my right hand is made of green cheese except when people examine it, that God ceases to exist every Wednesday afternoon, that cars feel thirsty when their tanks run low, or that cats think in German. I could make up hundreds of such interesting and irrefutable beliefs that no one would ever consider as a possible belief.

Religion is *not* a domain where anything goes, where any strange belief could appear and get transmitted from generation to generation. On the contrary, there is only a limited catalogue of possible supernatural beliefs. Even without knowing the details of religious systems in other cultures, we all know that some notions are far more widespread than others. The idea that there are invisible souls of dead people lurking around is a very common one; the notion that people's organs change position during the night is very rare. But both are equally irrefutable. So the problem, surely, is not just to explain how people can accept supernatural claims for which there is no strong evidence but also why they tend to represent and accept these particular supernatural claims rather than other possible ones. We should explain why they are so selective in the claims they adhere to.

Indeed, we should go even further and abandon the credulity-scenario altogether. Here is why: In this scenario, people relax ordinary standards of evidence for some reason. If you are against religion, you will say that this is because they are naturally credulous, or respectful of received authority, or too lazy to think for themselves, etc. If you are more sympathetic to religious beliefs, you will say that they open up their minds to wondrous truths beyond the reach of reason. But the point is that if you accept this account, you assume that people *first* open up their minds, as it were; and *then* let it be filled by whatever religious beliefs are held by the people who influence them at that particular time. This is often the way we think of religious adhesion. There is a gate-keeper in the mind that either allows or rejects visitors, that is, other people's concepts and beliefs. When the gate-keeper allows them in, these concepts and beliefs find a home in the mind and become the person's own beliefs and concepts.

Our present knowledge of mental processes suggests that this scenario is highly misleading. People receive all sorts of information from all sorts of sources. *All* this information has some effect on the mind. Whatever you hear and whatever you see is perceived, interpreted, explained, and recorded by the various inference systems I described above. Every bit of information is fodder for the mental machinery. But then some pieces of information produce the effects that we identify as "belief." That is, the person starts to recall them and use them to explain or interpret particular events; they may trigger specific emotions; they may strongly influence the person's behaviour. Note that I said *some* pieces of information, not all. This is where the selection occurs. In ways that a good psychology of religion should describe, it so happens that only some pieces of information trigger these effects, and not others; it also happens that the same piece of information will have these effects in some people but not others. So people do not have beliefs because they somehow made their minds receptive to belief and then acquired the material for belief. They have some beliefs because, among all the material they acquired, some of it triggered these particular effects.

A LIMITED CATALOGUE OF CONCEPTS

Do people know what their religious concepts are? This may seem an absurd question, but it is in fact an important question in the psychology of religion, whose true answer is probably in the negative. . . . For instance, psychologist Justin Barrett showed that Christians' concept of God was much more complex than the believers themselves assumed. Most Christians would describe their notion of God in terms of transcendence and extraordinary physical and mental characteristics. God is everywhere, attends to everything at the same time. However, subtle experimental tasks reveal that, when they are not reflecting upon their own beliefs, these same people use another concept of God, as a human-like agent with a particular viewpoint, a particular position and serial attention. God considers one problem and then another. Now that concept is mostly tacit. It drives people's thoughts about particular events, episodes of interaction with God, but it is not accessible to people as "their belief." In other words, people do not believe what they believe they believe.

A systematic investigation of these tacit concepts reveals that notions of religious agency, despite important cultural differences, are very similar the world over. There is a small repertoire of possible types of supernatural characters, many of whom are found in folktales and other minor cultural domains, though some of them belong to the important gods or spirits or ancestors of "religion." Most of these agents are explicitly defined as having counterintuitive physical or biological properties that violate general expectations about agents. They are sometimes undetectable, or prescient, or eternal. The way people represent such agents activates the enormous but inaccessible machinery of "theory of mind" and other mental systems that provide us with a representation of agents, their intentions and their beliefs. All this is inaccessible to conscious inspection and requires no social transmission. On the other hand, what is socially transmitted are the counterintuitive features: this one is omniscient, that one can go through walls, another one was born of a virgin, etc.

More generally, we observe that most supernatural and religious concepts belong to a short catalogue of possible types of templates, with a common structure. All these concepts are informed by very general assumptions from broad categories such as *person*, *living thing*, or *man-made object*. A spirit is a special kind of person, a magic wand a special kind of artifact, a talking tree a special kind of plant. Such notions combine (i) specific features that violate some default expectations for the domain with (ii) expectations held by default as true of the entire domain. For example, the familiar concept of a *ghost* combines (i) socially transmitted information about a physically counterintuitive person (disembodied, can go through walls, etc.), and (ii) spontaneous inferences afforded by the general person concept (the ghost perceives what happens, recalls what he or she perceived, forms beliefs on the basis of such perceptions, and intentions on the basis of beliefs).

These combinations of explicit violation and tacit inferences are culturally widespread and may constitute a memory optimum. Associations of this type are recalled better than more standard associations but also better than oddities that do not include domain-concept violations. The effect obtains regardless of exposure to a particular kind of supernatural beliefs, and it has been replicated in different cultures in Africa and Asia.

To sum up, we can explain human sensitivity to particular kinds of supernatural concepts as a by-product of the way human minds operate in ordinary, non-religious contexts. Because our assumptions about fundamental categories like *person, artifact, animal*, etc., are so entrenched, violations of these assumptions create salient and memorable concepts.

EXCHANGE, MORALITY, AND MISFORTUNE

We can understand other aspects of religious concepts as by-products of these ordinary, non-religious mental systems that organize our everyday experience. For instance, consider the fact that in all human cultures, a great deal of attention is focused, not so much on the characteristics of supernatural agents, as on their interaction with the living. This is visible in the constant association between moral judgments and supernatural agency, as well as in the treatment of misfortune and contingency.

Developmental research shows the early appearance and systematic organization of moral intuitions: a set of precise feelings evoked by the consideration of actual and possible courses of action. Although people often state that their moral rules are a consequence of the existence (or of the decrees) of supernatural agents, it is quite clear that such intuitions are present, independent of religious concepts. Moral intuitions appear long before children represent the powers of supernatural agents, they appear in the same way in cultures where no one is much interested in supernatural agents, and in similar ways regardless of what kind of supernatural agents are locally important. Indeed, it is difficult to find evidence that religious teachings have any effect on people's moral intuitions. Religious concepts do not change people's moral intuitions but frame these intuitions in terms that make them easier to think about. For instance, in most human groups supernatural agents are thought to be *interested parties* in people's interactions. Given this assumption, having the intuition that an action is wrong becomes having the expectation that a personalized agent disapproves of it. The social consequences of the latter way of representing the situation are much clearer to the agent, as they are handled by specialized mental systems for social interaction. This notion of gods and spirits as interested parties is far more salient in people's moral inferences than the notion of these agents as moral legislators or moral exemplars.

In the same way, the use of supernatural or religious explanations for misfortune may be a byproduct of a far more general tendency to see all salient occurrences in terms of social interaction. The ancestors can make you sick or ruin your plantations; God sends people various plagues. On the positive side, gods and spirits are also represented as protectors, guarantors of good crops, social harmony, etc. But why are supernatural agents construed as having such causal powers?

For these occurrences that largely escape control, people focus on the supernatural agents' feelings and intentions. The ancestors were angry, the gods demanded a sacrifice, or the god is just cruel and playful. But there is more to that. The way these reasons are expressed is, in a great majority of cases, supported by our *social exchange* intuitions. People focus on an agent's reasons for causing them harm, but note that these "reasons" always have to do with people's *interaction* with the agents in question. People refused to follow God's orders; they polluted a house against the ancestors' prescriptions; they had more wealth or good fortune than their God-decreed fate allocated them; and so on. All this supports what anthropologists have been saying for a long time on the basis of evidence gathered in the most various cultural environments: Misfortune is generally interpreted in *social* terms. But this familiar conclusion implies that the evolved cognitive resources people bring to the understanding of interaction should be crucial to their construal of misfortune.

Social interaction requires the operation of complex mental systems: to represent not just other people's beliefs and their intentions, but also the extent to which they can be trusted, the extent to which they find us trustworthy, how social exchange works, how to detect cheaters, how to build alliances, and so on. These mental systems are largely inaccessible, only their output is consciously represented. Now interaction with supernatural agents, through sacrifice, ritual, prayer, etc., is framed by those systems. Although the agents are said to be very special, the way people think about interaction with them is directly mapped from their interaction with actual people.

WHAT MAKES RELIGION "NATURAL"

For lack of space, I cannot pursue this list of the mental systems (usually activated in non-religious contexts) that sustain the salience and plausibility of religious notions. To be exhaustive, one should also mention the close association between ritual participation and group affiliation, the role of our coalitional thinking in creating religious identity, the specific role of death and dead bodies in religious thinking, and many other aspects of religion. Psychological investigation into these domains reveals the same organization described above. A variety of mental systems, functionally specialized for the treatment of particular (non-religious) domains of information, are activated by religious notions and norms, in such a way that these notions and norms become highly salient, easy to acquire, easy to remember and communicate, as well as intuitively plausible.

The lesson of the cognitive study of religion is that religion is rather "natural" in the sense that it consists of by-products of normal mental functioning. Each of the systems described here (a sense for social exchange, a specific mechanism for detecting animacy in surrounding objects, an intuitive fear of invisible contamination, a capacity for coalitional thinking, etc.) is the plausible result of selective pressures on cognitive organization. In other words, these capacities are the outcome of evolution by natural selection.

In other words, religious thought activates cognitive capacities that developed to handle non-religious information. In this sense, religion is very similar to music and very different from language. Every normal human being acquires a natural language and that language is extraordinarily similar to that of the surrounding group. It seems plausible that our capacity for language acquisition is an adaptation. By contrast, though all human beings can effortlessly recognize music and religious concepts, there are profound individual differences in the extent to which they enjoy music or adhere to religious concepts. The fact that some religious notions have been found in every human group does not mean that all human beings are naturally religious. Vast numbers of human beings do without it altogether, like for instance the majority of Europeans for several centuries.

Annotated Bibliography

Alexander, Richard D. 1987. *The Biology of Moral Systems*. New York: Aldine de Gruyter. Notable for its emphasis on "indirect reciprocity" as a central concept in understanding moral behavior and connecting it to evolutionary explanations of how behavior serves reproductive interests.

Aristotle. 1984. *The Complete Works of Aristotle: The Revised Oxford Translation*. Edited by Jonathan Barnes. Princeton: Princeton University Press. For naturalistic moral theories, the most important books are the *Nicomachean Ethics* and *Eudemian Ethics*.

Atran, Scott. 2002. *In Gods We Trust: The Evolutionary Landscape of Religion*. New York: Oxford University Press. A detailed presentation of Atran's cognitively based theory of religion. He examines rival theories in great detail, concluding that the human propensity to supernatural belief is an evolutionary by-product. One of the most solid attempts to scientifically explain religion that is available.

Aunger, Robert. 2002. *The Electric Meme: A New Theory of How We Think*. New York: The Free Press. Criticizes previous views of memes—self-replicating units of culture—and proposes a new model more closely based on brain and cognitive science. The best case so far for memes becoming a fruitful scientific idea.

Baggini, Julian. 2003. *Atheism: A Very Short Introduction*. New York: Oxford University Press. A brief, very accessible introduction that naturally lacks depth, but gives a good sense of how naturalism is central to modern nonbelief, and how naturalism is closely connected to science.

Baier, Kurt. 1995. *The Rational and the Moral Order: The Social Roots of Reason and Morality*. La Salle, IL: Open Court. Argues that morality can be based on rational calculations of self-interest. Although Baier makes many interesting arguments, his conclusion that rationality requires an egalitarian social order is strained.

Bailey, Lee W., and Jenny Yates. 1996. *The Near-Death Experience: A Reader*. New York: Routledge. Introductory essays representing many different points of view on NDEs.

Bainbridge, William Sims. 1997. *The Sociology of Religious Movements*. New York: Routledge. Mainly about new religions and their growth and change.

Barbour, Ian G. 2000. *When Science Meets Religion: Enemies, Strangers, or Partners?* New York: HarperSanFrancisco. A liberal theological argument for partnership between science and religion. Though its substance is dubious, Barbour's approach is important, as views like his have become very influential in framing the current "dialogue" between science and religion.

Behe, Michael J. 1996. *Darwin's Black Box: The Biochemical Challenge to Evolution*. New York: The Free Press. The leading biochemistry-based case for intelligent design creationism. Though this book was much criticized by mainstream scientists, the ID movement continues to defend its arguments.

Bishop, George. 2000. "Back to the Garden." *Public Perspective* 11(3): 21–23.

Blackmore, Susan. 1993. *Dying to Live: Near-Death Experiences*. Amherst, NY: Prometheus. Proposes a neurological hypothesis to explain reports of near-death experiences, arguing they are due to the peculiarities of a dying brain.

———. 1999. *The Meme Machine*. Oxford: Oxford University Press. The fullest statement of the imitation view of memetics: that memes, units of culture that independently replicate, are behaviors passed along by the unusual human capacity to imitate.

Block, Ned, Owen Flanagan, and Güven Güzeldere, eds. 1997. *The Nature of Consciousness: Philosophical Debates*. Cambridge: The MIT Press. Collects important philosophical papers and outlines the present contours of the philosophical debate over consciousness.

Boolos, George, John P. Burgess, and Richard C. Jeffrey. 2002. *Computability and Logic*. 4th ed. New York: Cambridge University Press. Introduction to basic concepts of theoretical computer science and mathematical logic.

Bowler, Peter J. 1988. *The Non-Darwinian Revolution: Reinterpreting a Historical Myth.* Baltimore: The Johns Hopkins University Press. Shows how while Darwin's theory was instrumental in convincing nineteenth-century scientists of the reality of common descent, natural selection as a mechanism was not widely accepted until well into the twentieth century. It also replies to objections that Darwin's theory was not original, or that it directly reflected political preoccupations of his time.

Boyer, Pascal. 2001. *Religion Explained: The Evolutionary Origins of Religious Thought.* New York: Basic. A weighty but readable introduction to the cognitive anthropological approach to explaining the origins and persistence of supernatural religion. Boyer illuminates commonalities in many religions, from orthodox Western theism to belief in ancestral spirits and witchcraft, and shows how the evolutionary history of human minds makes people predisposed to supernatural beliefs.

Bridgman, Percy W. 1947. "New Vistas for Intelligence." In *Physical Science and Human Values,* edited by E. P. Winger. Princeton, NJ: Princeton University Press.

Brooks, Daniel R., and E. O. Wiley. 1988. *Evolution as Entropy: Toward a Unified Theory of Biology.* 2nd ed. Chicago: University of Chicago Press. How biological evolution fits in with the second law of thermodynamics.

Brown, Warren S., Nancey Murphy, and H. Newton Malony, eds. 1998. *Whatever Happened to the Soul? Scientific and Theological Portraits of Human Nature.* Minneapolis, MN: Fortress Press. Defends a version of "nonreductive physicalism" concerning minds that includes "top-down causation," and argues its compatibility with liberal Christian theology.

Bruce, Steve. 1996. *Religion in the Modern World: From Cathedrals to Cults.* Oxford: Oxford University Press. A leading defender of the secularization thesis in the sociology of religion, Bruce argues that modern Western societies have been secularizing due to the changes brought on by modernity.

———. 1999. *Choice and Religion: A Critique of Rational Choice Theory.* Oxford: Oxford University Press. Argues that diversity and choice do not promote religiosity but undermine it, and that ethnic considerations have more to do with religious change than the structure of the religious marketplace.

———. 2002. *God Is Dead: Secularization in the West*. Malden, MA: Blackwell. Restates and defends the secularization thesis concerning Western Europe.

Büchner, Ludwig. 1884. *Force and Matter, or, Principles of the Natural Order of the Universe. With a System of Morality Based Thereupon.* Translated from the 15th German edition; 4th English edition. London: Asher and Co. A classic of nineteenth-century materialism.

Bunge, Mario. 2001. *Philosophy in Crisis: The Need for Reconstruction*. Amherst, NY: Prometheus. Bunge presents outlines of a comprehensive, scientific materialist view of everything from science and pseudoscience to neuroscience, social science, ethics, religion, and the nature of philosophy itself. He does not have room to fully argue for all of his judgments, which are often extremely critical, so this book is useful largely as a detailed summary of the views of an important thinker who takes modern science very seriously as a primary motivation for nonbelief.

Burton, Dan, and David Grandy. 2004. *Magic, Mystery, and Science: The Occult in Western Civilization*. Bloomington: Indiana University Press. A sympathetic history of the Western occult tradition, contrasted to scientific rationalism and orthodox religion alike. It is valuable for understanding the attraction and internal logic of occult and New Age views, which picture the universe as an organic, living whole permeated by mind and purpose.

Callahan, Tim. 1997. *Bible Prophecy: Failure or Fulfillment?* Altadena, CA: Millennium. A nonbelieving response to conservative Christian claims that fulfilled prophecy in the Bible demonstrates its divine inspiration.

Carrier, Richard. 2005. *Sense and Goodness without God: A Defense of Metaphysical Naturalism*. Bloomington, IN: Authorhouse. A comprehensive argument defending a godless, naturalistic view of the world. Though Carrier addresses many of the more philosophical issues concerning religion, his viewpoint is consistently linked to modern science. The book is also remarkably free of technical jargon, even when introducing complex and controversial ideas.

Chaison, Eric J. 2001. *Cosmic Evolution: The Rise of Complexity in Nature*. Cambridge: Harvard University Press. A physicist gives a view of evolution that encompasses change and the achievement of complexity at all levels throughout the universe. Though Chaison is reluctant to judge the truth of religious views, his scenario is completely and uncompromisingly naturalistic. Hence, supernatural agents would be out of place in the universe as envisioned by Chaison.

Chaitin, Gregory J. 1987. *Algorithmic Information Theory.* Cambridge: Cambridge University Press. Gives a rigorous mathematical definition of randomness as patternlessness.

Chesworth, Amanda, et al., eds. 2002. *Darwin Day Collection One: The Single Best Idea, Ever.* Albuquerque, NM: Tangled Bank. Contributions illustrate the tensions between separate spheres and conflict views of science and religion. Strictly scientific defenders of evolution tend to favor separate spheres.

Churchland, Paul M. 1996. *The Engine of Reason, The Seat of the Soul: A Philosophical Journey into the Brain.* Cambridge, MA: The MIT Press. A leading materialist philosopher of mind describes current brain research, arguing about its consequences for understanding consciousness and other aspects of the human mind. A good introduction to brain science.

Clark, Austen. 1993. *Sensory Qualities.* Oxford: Oxford University Press. A significant philosophical step toward a psychophysical understanding of qualia.

Cosmides, Leda, and John Tooby. 1994. "Beyond Intuition and Instinct Blindness: Toward an Evolutionarily Rigorous Cognitive Science." *Cognition* 50: 41–77.

Craig, William Lane, and Quentin Smith. 1993. *Theism, Atheism and Big Bang Cosmology.* Oxford: Clarendon. Craig, a conservative Christian philosopher, and Smith, a nonbelieving philosopher, debate whether the big bang was a moment of divine creation. Smith's discussion is more closely based on physics, and Craig relies too much on commonsense metaphysical intuitions.

Cupitt, Don. 2002. *Is Nothing Sacred? The Non-realist Philosophy of Religion.* New York: Fordham University Press. Radical theological essays defending a view of religion that does not depend on any supernatural beliefs, such as any literally existing God.

Damasio, Antonio. 1999. *The Feeling of What Happens: Body and Emotion in the Making of Consciousness.* San Diego, CA: Harcourt. A neurologist gestures toward a theory of how the brain gives rise to conscious awareness, based particularly on his work about emotions and their connection to body states. Though not a complete theory, work such as this demonstrates that naturalistic approaches are making progress on the problem of consciousness.

Danielson, Dennis Richard, ed. 2000. *The Book of the Cosmos: Imagining the Universe from Heraclitus to Hawking.* Cambridge, MA: Perseus. A good, nontechnical survey of thoughts about the universe from ancient philosophers to today's physical cosmologists. Particularly good in its focus on big, ambitious ideas concerning the universe as a whole.

Darwall, Stephen, Allan Gibbard, and Peter Railton. 1997. *Moral Discourse and Practice: Some Philosophical Approaches.* New York: Oxford University Press. A survey of current, secular moral philosophy.

Davies, Paul. 2004. "Multiverse Cosmological Models." *Modern Physics Letters A* 19(10): 727–743. A survey of multiple universe models that points out that they are plausible, and that they have matured to the degree of leading to testable predictions.

Dawkins, Richard. 1986. *The Blind Watchmaker: Why the Evidence of Evolution Reveals a Universe without Design.* New York: Norton. A classic semi-popular description and defense of evolution as a blind natural process.

———. 1989. *The Selfish Gene.* 2nd ed. Oxford: Oxford University Press. Defends a gene-centered view of evolution. Also notable for introducing the idea of "memes," replicators that are units of culture.

———. 1995. *River Out of Eden: A Darwinian View of Life.* New York: Basic. Emphasizes the purely natural, unguided nature of biological evolution.

———. 1998. *Unweaving the Rainbow: Science, Delusion and the Appetite for Wonder.* Boston: Houghton Mifflin. Upholds science and naturalism as a cultural and aesthetic ideal, replying to the charge that by cutting things apart and removing the mystery from nature, science deadens our appreciation of the world. A good example of the common contention of science-minded nonbelievers that even without gods, science itself can supply a sense of wonder and awe.

De Duve, Christian. 2002. *Life Evolving: Molecules, Mind, and Meaning.* New York: Oxford University Press. A comprehensive view of life, from chemical evolution and the origin of life through biological evolution and the brain. The Nobel laureate biologist emphasizes that nothing supernatural is needed to understand the world and ends with a somewhat unrealistic plea for religions to purge their beliefs of the supernatural, including a personal God.

Dembski, William A. 1999. *Intelligent Design: The Bridge between Science and Religion.* Downers Grove, IL: InterVarsity. A semipopular elaboration of Dembski's mathematical and information-based arguments against Darwinian evolution, together with theological and philosophical implications.

————. 2004. *The Design Revolution: Answering the Toughest Questions about Intelligent Design.* Downers Grove, IL: InterVarsity. Collects and clarifies much that has been said about intelligent design. The result is a theory that looks like it is of philosophical and theological interest; however, its scientific support remains practically nonexistent. Dembski's response to this weakness is to ignore scientific critics and declare victory.

Dembski, William A., and James M. Kushiner, eds. 2001. *Signs of Intelligence: Understanding Intelligent Design.* Grand Rapids: Brazos. Accessible collection of essays describing and defending the main themes of the recent anti-evolutionary movement called intelligent design.

Dembski, William A., and Michael Ruse, eds. 2004. *Debating Design: From Darwin to DNA.* Cambridge: Cambridge University Press. A collection of essays, largely by theologians and theology-minded scientists, on the theme of the argument from design. This volume brings together liberal, compatibilist theological positions on evolution together with anti-evolutionary views in the intelligent design camp. Strikingly, the liberals often flirt with intelligent-design-like ideas even as they avoid a direct challenge to modern science.

Dennett, Daniel C. 1991. *Consciousness Explained.* Boston: Little, Brown and Company. Although the title claims too much, Dennett goes some distance toward demystifying consciousness. A good example of a materialist, science-inspired approach to a traditional and religiously significant problem in philosophy.

————. 1995. *Darwin's Dangerous Idea: Evolution and the Meanings of Life.* New York: Simon and Schuster. Extends Darwinian evolution beyond biology, particularly using it in explaining how minds work. Presents Darwinian variation-and-selection as a "universal acid" eating away at all supernaturalistic, top-down pictures of the world. Though it sometimes gets lost in debates within evolutionary theory, this is an important book to get a picture of the central role of Darwinian thinking in today's naturalism.

————. 2003. *Freedom Evolves.* New York: Viking. Dennett applies his evolution-based, naturalistic view of human minds to the question of free will.

Denzler, Brenda. 2001. *The Lure of the Edge: Scientific Passions, Religious Beliefs, and the Pursuit of UFOs.* Berkeley: University of California Press. A study of UFOlogy from a cultural- and religious-studies point of view.

Duncan, Otis Dudley. 2004. "The Rise of the Nones: A Paleostatistical Inquiry, Part 2." *Free Inquiry* 24(2): 29–31.

Edelman, Gerald M. 2004. *Wider Than the Sky: The Phenomenal Gift of Consciousness.* New Haven: Yale University Press. Edelman presents a more accessible description of the theory expounded in Edelman and Tononi 2000, and speculates further on how the features of human consciousness that go beyond primary awareness can be achieved by the brain.

Edelman, Gerald M., and Giulio Tononi. 2000. *A Universe of Consciousness: How Matter Becomes Imagination.* New York: Basic. Lays out a well-developed theory of how consciousness can be identified with a certain kind of brain state. While not the last word, this book illustrates how today's brain science is making real progress in understanding the mind.

Edis, Taner. 1998a. "How Gödel's Theorem Supports the Possibility of Machine Intelligence." *Minds and Machines* 8: 251–262.

———. 1998b. "Taking Creationism Seriously." *Skeptic* 6(2): 56–65.

———. 2002. *The Ghost in the Universe: God in Light of Modern Science.* Amherst, NY: Prometheus. A defense of a comprehensively naturalist view of the world, critiquing God and other supernatural ideas from the point of view of contemporary science. While defending a science-minded nonbelief, the book also emphasizes the continuity between science and philosophy.

———. 2003a. "Flipping a Quantum Coin." *Free Inquiry* 23(2): 60.

———. 2003b. "A World Designed by God: Science and Creationism in Contemporary Islam." In *Science and Religion: Are They Compatible?* edited by Paul Kurtz. Amherst, NY: Prometheus.

———. 2004a. "Exorcizing All the Ghosts." *The Skeptical Inquirer* 28(2): 35–38, 48.

———. 2004b. "Chance and Necessity—and Intelligent Design?" In *Why Intelligent Design Fails: A Scientific Critique of the New Creationism*, edited by Matt Young and Taner Edis. New Brunswick, NJ: Rutgers University Press.

Eisenach, Eldon J. 2000. *The Next Religious Establishment: National Identity and Political Theology in Post-Protestant America.* Lanham: Rowman and Littlefield. Describes how even with separation of church and state, American religion has had a series of informal establishments, and how this has been crucial for Americans' sense of national identity. Useful in understanding the complexities of the church/state debate in the United States, and seeing why nonbelievers' concerns are politically impotent.

Eller, David. 2004. *Natural Atheism*. Cranford, NJ: American Atheist Press. A cultural anthropologist argues for atheism. Although concentrating on familiar philosophical matters, he also includes discussions on science and religion, and explains how awareness of cultural diversity undermines religious convictions.

Eve, Raymond A., and Francis B. Harrold. 1991. *The Creationist Movement in Modern America*. Boston: Twayne. A sociological study of creationism. Especially useful in pointing out how creationists are typically very positive about technology and are attracted to "scientific" creationism because of their need to reconcile science and religion.

Fakhry, Majid. 2004. *A History of Islamic Philosophy*. 3rd ed. New York: Columbia University Press. A wide-ranging historical survey of Islamic philosophy, including rationalist and modernist trends as well as orthodox views.

Flamm, Bruce. 2004. "The Columbia University 'Miracle' Study: Flawed and Fraud." *Skeptical Inquirer* 28(5): 25–31.

Flanagan, Owen. 2002. *The Problem of the Soul: Two Visions of Mind and How to Reconcile Them*. New York: Basic. Defends a scientific, materialist view of mind, and explores its implications for questions about free will, the nature of persons, and ethics.

Forman, Robert K. C., ed. 1990. *The Problem of Pure Consciousness: Mysticism and Philosophy*. Oxford: Oxford University Press. Contributions include interesting philosophical responses to naturalistic, brain-based explanations of mystical experience and their skeptical implications concerning the evidential value of mystical experience for the divine.

Forrest, Barbara, and Paul R. Gross. 2004. *Creationism's Trojan Horse: The Wedge of Intelligent Design*. New York: Oxford University Press. An in-depth examination of the social and political aspects of the intelligent design movement. Helps set the relationship of science, conservative religion, and nonbelief in context.

Fox, Mark. 2003. *Religion, Spirituality and the Near-Death Experience*. New York: Routledge. Although Fox has considerable sympathy toward religious and dualist interpretations of NDEs, he extensively documents how the case for a dualist understanding of NDEs has not been made, and that NDE interpretations are socially and culturally constructed.

Freud, Sigmund. 1928. *The Future of an Illusion*. Translated by W. D. Robson-Scott. New York: H. Liveright. Freud's view of religion is a classic of nonbelief, with interesting insights as well as psychoanalytic ideas that have entirely lost scientific credibility today.

Fuller, Robert C. 2001. *Spiritual but not Religious: Understanding Unchurched America*. New York: Oxford University Press. Explores the history and present of individualist religiosity in the United States, which can be similar to nonbelieving movements when criticizing the dogmatism and moral authoritarianism of organized religion, and yet retains a strong commitment to supernatural beliefs.

Garner, Richard. 1994. *Beyond Morality*. Philadelphia: Temple University Press. Defends an "amoralist" view, according to which the notion of objective moral facts is an error.

Gasperini, Maurizio, and Gabriele Veneziano. 2003. "The Pre-Big Bang Scenario in String Cosmology." *Physics Reports* 373(1–2): 1–212. For a nontechnical description, see Gabriele Veneziano, "The Myth of the Beginning of Time." 2004. *Scientific American* 290(5): 56–65.

Gillett, Carl, and Barry Loewer, eds. 2001. *Physicalism and Its Discontents*. Cambridge: Cambridge University Press. Technical philosophical essays defending and attacking physicalism, the strongest modern version of materialism. It emphasizes debates about reductionism and conceptual issues in the philosophy of mind, rather than actual physics.

Gishlick, Alan D. 2004. "Evolutionary Paths to Irreducible Systems: The Avian Flight Apparatus." In *Why Intelligent Design Fails: A Scientific Critique of the New Creationism*, edited by Matt Young and Taner Edis. New Brunswick, NJ: Rutgers University Press.

Goldsmith, Donald, and Tobias C. Owen. 2001. *The Search for Life in the Universe*. 3rd ed. Sausalito, CA: University Science Books. A college textbook that gives a relatively nontechnical overview of bioastronomy and the current thinking about the abundance of life and intelligent life in the universe.

Gonzalez, Guillermo, and Jay Wesley Richards. 2004. *The Privileged Planet: How Our Place in the Cosmos Is Designed for Discovery*. Washington, DC: Regnery. An astronomical contribution to the intelligent design movement. Argues that Earth is unique in the universe in its intelligent life-sustaining properties.

Goswami, Amit. 2001. *Physics of the Soul: The Quantum Book of Living, Dying, Reincarnation and Immortality*. Charlottesville, VA: Hampton Roads. Goswami, a legitimate physicist, ventures far outside of the mainstream in attempting to find a basis for the soul and reincarnation in quantum mechanics. A science-based apologetic for Eastern and New Age spirituality.

Gould, Stephen Jay. 1980. *The Panda's Thumb: More Reflections in Natural History*. New York: Norton. Collection of essays on the topic of evolution.

————. 1989. *Wonderful Life: The Burgess Shale and the Nature of History*. New York: W. W. Norton. Gould argues that evolution produces an extremely contingent, unpredictable history. The degree to which this is correct is a matter of dispute; Gould did not adequately account for the connections of many Burgess fossils with known phyla.

————. 1996. *Full House: The Spread of Excellence from Plato to Darwin*. New York: Three Rivers. Gould argues that not only is evolution not inherently progressive, but that increasing complexity in some lineages is only due to the spread over time of the tail end of a distribution.

————. 1999. *Rocks of Ages: Science and Religion in the Fullness of Life*. New York: Ballantine. Defines and defends a version of separate spheres called "non-overlapping magisteria." Science and religion are supposed to have their own separate concerns and competencies, the world of facts and the realm of values.

Grant, Edward. 2004. *Science and Religion, 400 B.C. to A.D. 1550: From Aristotle to Copernicus*. Westport, CT: Greenwood Press.

Greene, Brian. 1999. *The Elegant Universe: Superstrings, Hidden Dimensions, and the Quest for the Ultimate Theory*. New York: W. W. Norton. An accessible introduction to the basic ideas behind string theory, a possible path to achieving a theory of quantum gravity.

Greene, John C. 1959. *The Death of Adam: Evolution and its Impact on Western Thought*. Ames: The Iowa State University Press. A classic intellectual history of Darwinian evolution.

Gregersen, Niels Henrik, ed. 2003. *From Complexity to Life: On the Emergence of Life and Meaning*. New York: Oxford University Press. Combines ideas about chaos theory, self-organization, and emergence with theological speculation on how all this reveals a world impelled to increasing complexity by divine design.

Griffin, David Ray. 1997. *Parapsychology, Philosophy, and Spirituality: A Postmodern Exploration*. Albany: SUNY Press. A liberal theologian argues for the reality of psychic powers and paranormal events, and uses them to support a religious view of the world. Griffin defends an unconventional notion of "naturalism," in the sense that the natural and supernatural shade gradually into each other.

Grim, Patrick, ed. 1990. *Philosophy of Science and the Occult*. Albany: SUNY Press. Philosophically oriented critiques and defenses of paranormal claims.

Gross, Paul R., Norman Levitt, and Martin W. Lewis, eds. 1996. *The Flight from Science and Reason*. New York: The New York Academy of Sciences. Essays defending science against criticism from philosophical, postmodernist, and religious sources.

Guth, Alan H. 1997. *The Inflationary Universe: The Quest for a New Theory of Cosmic Origins*. Reading, MA: Addison-Wesley. The history and implications of inflationary cosmology described by the physicist who originated the idea.

Guthrie, Stewart Elliott. 1993. *Faces in the Clouds: A New Theory of Religion*. New York: Oxford University Press. An influential book arguing that religion is based on the human tendency to anthropomorphize, to infer purposes behind complex and ambiguous stimuli.

Habermas, Gary R., and J. P. Moreland. 1998. *Beyond Death: Exploring the Evidence for Immortality*. Wheaton, IL: Crossway. A defense of a conventionally theistic view of the mind, based on philosophical arguments for mind-brain dualism, bolstered by paranormal evidence such as near-death experiences, and more religious arguments for the reality of a soul.

Hamer, Dean. 2004. *The God Gene: How Faith Is Hardwired into Our Genes*. New York: Doubleday. A geneticist makes a weak case for not only a genetic basis for religiosity but a single gene being implicated. Hamer presents some interesting ideas, but overreaches his evidence.

Harman, Gilbert. 2000. *Explaining Value and Other Essays in Moral Philosophy*. New York: Oxford University Press. Defends a relativist view of morality. The chapter "Is There a Single True Morality?" is particularly relevant, as it brings out some of the connections between naturalism and moral relativism.

Harris, Sam. 2004. *The End of Faith: Religion, Terror, and the Future of Reason*. New York: W. W. Norton. An example of the kind of ill-informed, hysterical antireligiosity that too often attracts nonbelievers. Harris denounces all faith, particularly Islam, although his own views have a considerable element of mysticism and sympathy to the paranormal.

Hasker, William. 1999. *The Emergent Self*. Ithaca, NY: Cornell University Press. Defends a religion-friendly "emergent dualism" as a philosophy of mind.

Haught, John F. 2000. *God after Darwin: A Theology of Evolution*. Boulder, CO: Westview. Representative of liberal theological efforts to affirm modern biology but still argue that God is behind evolution. Some of Haught's ideas on how divine creativity is manifested in information are similar to those that surface in the intelligent design movement, but they are never made quite as concrete.

————. 2003. *Deeper Than Darwin: The Prospect for Religion in an Age of Evolution*. Boulder, CO: Westview. Continues the themes found in Haught 2000, with sharper philosophical attacks on materialist views of evolution, plus a critique of intelligent design as insufficiently "deep" in its theology.

Hawking, Stephen W. 2001. *The Universe in a Nutshell*. New York: Bantam. A well-known cosmologist describes current ideas in physical cosmology in a relatively nontechnical fashion.

Hecht, Jennifer Michael. 2003. *Doubt: A History*. New York: HarperSanFrancisco. A comprehensive and accessible survey of religious doubt. Hecht examines doubt within religious traditions as well as outright dissent and gives equal weight to Far Eastern and Indian versions of nonbelief. Due to her very broad approach, however, the sort of science-minded nonbelief that is the focus of this book is not addressed in depth.

Huff, Toby E. 2003. *The Rise of Early Modern Science: Islam, China, and the West*. 2nd ed. New York: Cambridge University Press. Asks why modern science emerged in the Christian West, rather than in those cultures that were intellectually advanced at the time, particularly Islam and China. Huff points to various historical accidents—plus institutional developments that allowed fields such as the law to develop relatively free of religious constraints—that prepared the way for independent inquiry and science.

Hume, David. 1992. *Writings on Religion*. Edited by Antony Flew. La Salle, IL: Open Court. Hume's most important and historically important works on religion, including *Dialogues concerning Natural Religion*, posthumously published in 1779. Though primarily philosophical in nature, Hume's arguments have inspired science-minded skepticism about religion as well.

Humphrey, Nicholas. 1996. *Leaps of Faith: Science, Miracles, and the Search for Supernatural Consolation*. New York: Basic. By a psychologist, this book contains some of the most interesting arguments against the existence of psychic powers.

Humphreys, D. Russell. 1994. *Starlight and Time: Solving the Puzzle of Distant Starlight in a Young Universe*. Green Forest, AR: Master Books. A young-earth creationist attempt to reconcile a few-thousand-year-old age for the universe and the vast size of the universe by an impressive exercise in pseudophysics.

Hyman, Ray. 1996. "Cold Reading: How to Convince Strangers You Know All about Them." In *The Outer Edge: Classic Investigations of the Paranormal*, edited by Joe Nickell, Barry Karr, and Tom Genoni. Amherst, NY: CSICOP.

Joshi, S. T. 2003. *God's Defenders: What They Believe and Why They Are Wrong*. Amherst, NY: Prometheus. An example of an atheist polemic, attacking popular rather than scholarly defenses of God. Representative of many popular arguments concerning God.

Katz, Steven T., ed. 1983. *Mysticism and Religious Traditions*. New York: Oxford University Press. Makes a case for the cultural construction of mystical experience. Katz's own essay, "The 'Conservative' Character of Mystical Experience," is especially interesting.

Kellehear, Allan. 1996. *Experiences Near Death: Beyond Medicine and Religion*. New York: Oxford University Press. Demonstrates that the social and cultural aspects of near death experiences must be taken into account; no single explanation like a simple medical phenomenon can account for NDEs, which include a wide variety of experiences.

Kirk, G. S., and J. E. Raven. 1962. *The Presocratic Philosophers: A Critical History with a Selection of Texts*. Cambridge: Cambridge University Press.

Klee, Robert. 2002. "The Revenge of Pythagoras: How a Mathematical Sharp Practice Undermines the Contemporary Design Argument in Astrophysical Cosmology." *British Journal for the Philosophy of Science* 53: 331–354.

Kurtz, Paul, ed. 1985. *A Skeptic's Handbook of Parapsychology*. Amherst, NY: Prometheus. Though not up-to-date on developments since the 1980s, this collection touches on most of the common skeptical criticisms of parapsychology and includes a number of historically significant papers.

————. 1989. *Eupraxophy: Living without Religion.* Amherst, NY: Prometheus. By a leading "secular humanist" philosopher, it explains and defends a nonbelieving approach to life and morality.

————. 1991. *The Transcendental Temptation: A Critique of Religion and the Paranormal.* Buffalo, NY: Prometheus. A comprehensive defense of a skeptical attitude toward supernatural beliefs. Notable in its treatment of religious and paranormal beliefs in the same way.

————, ed. 2001. *Skeptical Odysseys: Personal Accounts by the World's Leading Paranormal Inquirers.* Amherst, NY: Prometheus. The personal essays included give considerable insight into the world of paranormal skepticism, as well as reproducing some intriguing arguments against paranormal and supernatural beliefs of all sorts.

————, ed. 2003. *Science and Religion: Are They Compatible?* Amherst, NY: Prometheus. The contributions are mostly by skeptics who doubt that science and religion are compatible.

Lahav, Noam. 1999. *Biogenesis: Theories of Life's Origin.* New York: Oxford University Press. A technical survey of current ideas, knowledge, and uncertainties concerning chemical evolution and the origins of life.

Larson, Edward J., and Larry Witham. 1998. "Leading Scientists Still Reject God." *Nature* 394: 313.

Le Poidevin, Robin. 1996. *Arguing for Atheism: An Introduction to the Philosophy of Religion.* New York: Routledge. A good recent introduction to atheist philosophy, particularly valuable in addressing sophisticated cosmological arguments.

Lestienne, Rémy. 1998. *The Creative Power of Chance.* Urbana: University of Illinois Press. A forceful argument for how, in physics and biology, chance is fundamental and crucial for achieving real creativity.

Lewis, James R., ed. 1995. *The Gods Have Landed: New Religions from Outer Space.* Albany: SUNY Press. Essays on the religious aspects of UFO experiences and beliefs. Believers interpret their strange experiences in otherworldly terms, while social and behavioral scientists make sense of the phenomena within the natural world.

Lewis, James R., and J. Gordon Melton, eds. 1992. *Perspectives on the New Age.* Albany: SUNY Press. A collection of scholarly essays examining various aspects of the New Age movement. Useful for getting beyond the superficial images of the New Age found in popular media.

Lewy, Guenter. 1996. *Why America Needs Religion: Secular Morality and Its Discontents.* Grand Rapids, MI: William B. Eerdmans. A politically conservative defense of religion for pragmatic secular reasons.

Livingstone, David N. 1987. *Darwin's Forgotten Defenders: The Encounter between Evangelical Theology and Evolutionary Thought.* Grand Rapids, MI: Eerdmans. After Darwin's theory was published, many evangelicals accepted evolution, provided it was interpreted as a progressive development guided by God.

Loftus, Elizabeth, F., and Gary L. Wells, eds. 1984. *Eyewitness Testimony: Psychological Perspectives.* New York: Cambridge University Press. On the fallibility of eyewitness testimony. Psychological studies find they are unreliable, and that there are many factors that distort memory. So testimony is generally not worth much without independent corroborating evidence. Scientific skeptics therefore distrust eyewitness testimony for paranormal and miraculous events.

Lucretius. 1995. *On the Nature of Things: De Rerum Natura.* Translated by Anthony M. Esolen. Baltimore: The Johns Hopkins University Press. Esolen's translation of the atomist, materialist classic of antiquity.

Lyons, William. 2001. *Matters of the Mind.* New York: Routledge. A survey of philosophical thought concerning the mind, particularly in the twentieth century. An excellent introduction to the debate.

Mack, John E. 1994. *Abduction: Human Encounters with Aliens.* New York: Charles Scribner's Sons. A Harvard psychiatrist argues that UFO abduction experiences are due to real alien encounters, and that the materialist science of today disregards them for illegitimate reasons.

Mackie, J. L. 1982. *The Miracle of Theism: Arguments for and against the Existence of God.* Oxford: Oxford University Press. A classic representative of the philosophical style of religious nonbelief.

Manson, Neil A., ed. 2003. *God and Design: The Teleological Argument and Modern Science.* New York: Routledge. A survey of the main current arguments concerning whether the universe was designed by a theistic God. Includes proponents and critics of the intelligent design movement.

Marks, David. 2000. *The Psychology of the Psychic.* 2nd ed. Amherst, NY: Prometheus. A critical survey of prominent parapsychological claims, including both laboratory-based results and the performances of psychic superstars such as Uri Geller.

Martin, Michael. 1990. *Atheism: A Philosophical Justification.* Philadelphia: Temple University Press. A leading example of a defense of atheism from an analytic philosophical point of view.

————. 2002. *Atheism, Morality, and Meaning.* Amherst, NY: Prometheus. Includes an argument for the existence of fully objective moral facts, from an atheistic philosophical point of view. Martin's defense of objective morality for the nonbeliever is not entirely convincing, and his approach is notable for ignoring what the sciences have to say about the nature of moral perception and behavior.

May, Larry, Marilyn Friedman, and Andy Clark, eds. 1996. *Mind and Morals: Essays on Ethics and Cognitive Science.* Cambridge: The MIT Press. Differing philosophical viewpoints on the implications of science, particularly brain and cognitive science, where morality is concerned.

Mayr, Ernst. 1991. *One Long Argument: Charles Darwin and the Genesis of Modern Evolutionary Thought.* Cambridge: Harvard University Press. An important source on the philosophy of biology and the history of evolutionary ideas by one of the most important thinkers about evolution.

McClenon, James. 1994. *Wondrous Events: Foundations of Religious Belief.* Philadelphia: University of Pennsylvania Press. A sociologist argues that the supernatural beliefs of all cultures are based on universal human experiences, that these fall into categories explored by parapsychologists, and that there probably is good evidence for the reality of such supernatural powers.

McCutcheon, Russell T. 2001. *Critics Not Caretakers: Redescribing the Public Study of Religion.* Albany: SUNY Press. Argues that religion must be explained as a social and historical phenomenon, unlike the current theological approach in academic religious studies.

McGrath, Alister E. 2004. *The Twilight of Atheism: The Rise and Fall of Disbelief in the Modern World.* New York: Doubleday. McGrath, a theologian, points out the resurgence of religion and observes that atheism appears to have begun to decline into irrelevance. He is not quite convincing when discussing nonbelief in intellectual circles, and his views on science and religion are simplistic. He argues very well, however, that the public, political appeal of atheism has declined considerably, tracing how nonbelief is no longer seen as promising freedom.

McKinsey, C. Dennis. 2000. *Biblical Errancy: A Reference Guide.* Amherst, NY: Prometheus. A representative of the genre of nonbelieving literature focusing on errors and internal contradictions in the Bible. It includes scientific mistakes, but devotes comparatively little space to them.

Melnyk, Andrew. 2003. *A Physicalist Manifesto: Thoroughly Modern Materialism.* New York: Cambridge University Press. A detailed argument that everything is physical, in the sense that everything that we know exists are functional types that are physically realized. A very science-aware work of philosophy, which captures many of the insights leading scientists to physicalist views.

Meyer, Alden. 2004. "Bringing Science Back to the People." *Catalyst: The Magazine of the Union of Concerned Scientists* 3(1): 2–6.

Monod, Jacques. 1971. *Chance and Necessity: An Essay on the Natural Philosophy of Modern Biology.* New York: Knopf. An influential and much-criticized presentation of evolution's centrality to a naturalistic view of the world. Monod's thesis, that all we see can be explained by combinations of mechanisms and randomness, remains a strong motivation for today's science-based naturalism and physicalism.

Morris, Henry M. 1985. *Scientific Creationism.* El Cajon, CA: Master Books. The most authoritative text about the U.S. Protestant version of "scientific" young-earth creationism. Particularly notable for its wide variety of arguments that attempt to show the earth is a few thousand years old, although none are taken seriously by mainstream science.

———. 1993. "Dragons in Paradise." *Impact* 241: 1–4.

Morris, Simon Conway. 2003. *Life's Solution: Inevitable Humans in a Lonely Universe.* Cambridge: Cambridge University Press. A biologist controversially argues that evolution inevitably proceeds toward producing humans, and that evolution has an inherent direction and religious significance.

Moseley, James W., and Karl T. Pflock. 2002. *Shockingly Close to the Truth: Confessions of a Grave-robbing Ufologist.* Amherst, NY: Prometheus. A memoir about the UFO subculture.

Musgrave, Ian. 2004. "Evolution of the Bacterial Flagellum." In *Why Intelligent Design Fails: A Scientific Critique of the New Creationism,* edited by Matt Young and Taner Edis. New Brunswick, NJ: Rutgers University Press.

Nasr, Seyyed Hossein. 1989. *Knowledge and the Sacred.* Albany: SUNY Press. Writing from the perspective of the Muslim mystical and philosophical tradition, Nasr wants to subordinate natural science to religious knowledge. He also denounces the theory of evolution.

Newberg, Andrew, Eugene d'Aquili, and Vince Rause. 2002. *Why God Won't Go Away: Brain Science and the Biology of Belief.* New York: Ballantine. A strange combination of hard neurobiology and speculative theology. The opening chapters strongly suggests that certain religious experiences are artifacts of brain structure; the rest of the work attempts to develop a "neurotheology" to try to bring God back into the picture.

Nickell, Joe. 1993. *Looking for a Miracle: Weeping Icons, Relics, Stigmata, Visions and Healing Cures.* Amherst, NY: Prometheus. A skeptical investigator of paranormal claims debunks religious miracle claims, particularly those within the Catholic tradition.

Nickell, Joe, Barry Karr, and Tom Genoni, eds. 1996. *The Outer Edge: Classic Investigations of the Paranormal.* Amherst, NY: CSICOP. Skeptical essays on paranormal and fringe-science claims, arranged for use as a college textbook in critical thinking courses.

Norris, Pippa, and Ronald Inglehart. 2004. *Sacred and Secular: Religion and Politics Worldwide.* New York: Cambridge University Press. Includes up-to-date numbers on the prevalence of nonbelief in diverse populations across the globe. The authors also argue that religiosity declines in societies that enjoy a high degree of food security, good public health, accessible housing, and so forth.

Numbers, Ronald L. 1992. *The Creationists: The Evolution of Scientific Creationism.* Berkeley: University of California Press. An authoritative history of the "creation-science" movement.

OECD. 2004. *OECD Science, Technology, and Industry Outlook.* Paris: Organisation for Economic Co-operation and Development.

Olson, Richard G. 2004. *Science and Religion, 1450–1900: From Copernicus to Darwin.* Westport, CT: Greenwood Press.

Oppenheim, Janet. 1985. *The Other World: Spiritualism and Psychical Research in England, 1850–1914.* Cambridge: Cambridge University Press. A good historical overview of the ideas and motivations involved in early British psychical research.

Otto, Rudolf. 1923. *The Idea of the Holy.* London: Oxford University Press. A classic definition and defense of the idea of approaching God through mystical and religious experience.

Paley, William. 1802. *Natural Theology, or, Evidences of the Existence and Attributes of the Deity: Collected from the Appearances of Nature.* Philadelphia:

John Morgan. Possibly the most influential work of British natural theology; it is still used as a defining example of attempts to argue from
nature to the reality of a God.

Palmer, Susan J. 2004. *Aliens Adored: Rael's UFO Religion*. New Brunswick, NJ:
Rutgers University Press. A close study of the Raëlian UFO religion,
which replaces supernatural gods with space aliens.

Parsons, Keith M., ed. 2003. *The Science Wars: Debating Scientific Knowledge and
Technology*. Amherst, NY: Prometheus. A collection of articles largely
devoted to presenting postmodern critiques of science and responses
by defenders of science. Also includes a section on conservative critiques of science. A good introduction to the issues.

Peebles, Curtis. 1994. *Watch the Skies! A Chronicle of the Flying Saucer Myth*.
Washington, DC: Smithsonian Institution Press. An authoritative and
skeptical history of UFO-related beliefs.

Pennock, Robert T. 1999. *Tower of Babel: The Evidence against the New Creationism*. Cambridge, MA: The MIT Press. A critique of intelligent design
creationism, developed from a scientific and liberal theological point
of view. Includes a vigorous but flawed defense of "methodological
naturalism" as an essential feature of science.

————, ed. 2001. *Intelligent Design Creationism and Its Critics: Philosophical, Theological and Scientific Perspectives*. Cambridge, MA: The MIT Press.
Represents the technical arguments of both ID supporters and critics.
Comprehensive, but puts its emphasis on philosophical and theological arguments rather than scientific concerns.

Penrose, Roger. 1994. *Shadows of the Mind: A Search for the Missing Science of
Consciousness*. Oxford: Oxford University Press. An eminent mathematical physicist defends a quasi-mystical, Platonic view of consciousness. Notable for its argument based on Gödelian considerations
and trying to tie it to speculations in quantum gravity. Penrose's views
have found very little support.

Perakh, Mark. 2004. *Unintelligent Design*. Amherst, NY: Prometheus. A strong
critique of intelligent design, plus critical examinations of the many
popular attempts to claim religion finds support from modern physics.
Especially useful in its focus on the poor physics of certain Jewish
apologetic strategies.

Pigliucci, Massimo. 2000. *Tales of the Rational: Skeptical Essays about Nature and
Science*. Atlanta, GA: Freethought Press. A biologist interested in philosophy argues against creationism and religion in general.

Pinker, Steven. 2002. *The Blank Slate: The Modern Denial of Human Nature.* New York: Penguin. A semipopular exploration of what might be innate in human nature. It includes an accessible description and defense of the modular structure of human minds.

Plotkin, Henry. 1998. *Evolution in Mind: An Introduction to Evolutionary Psychology.* Cambridge, MA: Harvard University Press. The basic ideas of evolutionary psychology, which seeks adaptive explanations for common features of the human mind.

Polkinghorne, John. 1998. *Belief in God in an Age of Science.* New Haven, CT: Yale University Press. A physicist who became an Anglican priest explains how he brings modern science and Christianity into harmony. Notable for referring to some concepts from physics that might allow for divine action, but not for developing any concrete proposals.

Price, Huw. 1996. *Time's Arrow and Archimedes' Point: New Directions for the Physics of Time.* New York: Oxford University Press. How the microscopic time reversibility of physics connects to macroscopic reversibility in thermodynamic and cosmological contexts. Includes an interesting approach to quantum mechanics.

Price, Robert M, and Jeffery Jay Lowder, eds. 2005. *The Empty Tomb: Jesus beyond the Grave.* Amherst, NY: Prometheus. A response, by nonbelieving scholars, to current evangelical Christian defenses of the historicity of the resurrection of Jesus.

Pyysiäinen, Ilkka. 2004. *Magic, Miracles, and Religion: A Scientist's Perspective.* Walnut Creek, CA: AltaMira. Collected papers criticizing philosophical attempts to protect religion from scientific criticism, and defending aspects of a cognitive, science-based understanding of religious belief. This book is especially valuable in addressing the sophisticated theological approach to religion, as well as more obviously supernatural beliefs.

Rachels, James. 1991. *Created from Animals: The Moral Implications of Darwinism.* Oxford: Oxford University Press. This work argues that Darwinian evolution provides a stronger critique of the classical argument from design than David Hume's philosophical approach, and that Darwinism strains religious belief. Rachels also explores the moral implications of evolution, defending animal rights.

Radin, Dean. 1997. *The Conscious Universe: The Scientific Truth of Psychic Phenomena.* New York: HarperEdge. A leading parapsychologist argues that there is ample experimental evidence for the reality of psychic powers, that skeptics are misguided, and that a view of the world similar to that presented by Eastern and mystical religious traditions may be correct.

Randi, James. 1987. *The Faith Healers*. Amherst, NY: Prometheus. Contains sections on the history of faith healing, but the strength of the book is its direct examinations and exposés of particular faith healing claims. Randi's debunking of Peter Popoff is well known, and most of his examples are from the evangelical subculture. Not a comprehensive survey, but a good examination of real-world faith healing.

Rawls, John. 1999. *A Theory of Justice*. Rev. ed. Cambridge, MA: Harvard University Press. A classic of political philosophy.

Richerson, Peter J., and Robert Boyd. 2005. *Not by Genes Alone: How Culture Transformed Human Evolution*. Chicago: University of Chicago Press. Explores current ideas about culture-gene coevolution and the biological basis for culture as a uniquely human adaptation.

Roof, Wade Clark. 1999. *Spiritual Marketplace: Baby Boomers and the Remaking of American Religion*. Princeton, NJ: Princeton University Press. Describes the seeker and "quest culture" that has taken hold in U.S. religion, and the individualism inherent even in the way Americans shop for religious community. Illuminating, although Roof overextends the market metaphor.

Ruse, Michael. 2001. *Can a Darwinian Be a Christian: The Relationship between Science and Religion*. New York: Oxford University Press. A statement of what has become the conventional wisdom, that science and religion can comfortably occupy separate spheres. Ruse thinks a Darwinian can be a Christian, but the flexibility he perceives in theology is hard to distinguish from obscurantism.

Russell, Bertrand. 1935. *Religion and Science*. New York: Henry Holt. Although very out of date in many of the details of the science to which it refers, this book continues to be a good illustration of some of the basic attitudes and approaches of science-minded nonbelievers.

———. 1961. *History of Western Philosophy*. London: George Allen and Unwin. A classic, this is still a very good introduction to the philosophical context for religious doubt and its connection to science, especially since the philosophy of religion has not changed greatly within the last century.

Sagan, Carl. 1996. *The Demon-Haunted World: Science as a Candle in the Dark*.

New York: Random House. A noted scientist and skeptic sympatheti-
cally argues against paranormal beliefs and holds up science as a noble
human enterprise.

Saler, Benson, Charles A. Ziegler, and Charles B. Moore. 1997. *UFO Crash at
Roswell: The Genesis of a Modern Myth*. Washington, DC: Smithsonian
Institution Press. A thorough examination of the Roswell flying saucer
crash myth and the real events behind it. The chapter "Roswell and
Religion" is particularly interesting.

Sampson, Wallace, and Lewis Vaughn. 2000. *Science Meets Alternative Medicine: What the Evidence Says about Unconventional Treatments*. Amherst,
NY: Prometheus. Skeptical scientists and physicians examine a wide
range of claims concerning alternative medicine.

Scharfstein, Ben-Ami. 1973. *Mystical Experience*. Oxford: Blackwell. A classic
critique of the claims of mystics to be able to establish claims about ultimate reality through direct experience.

Schick, Jr. Theodore, and Lewis Vaughn. 2005. *How to Think about Weird
Things: Critical Thinking for a New Age*. 4th ed. Boston: McGraw-Hill. A
textbook for critical thinking courses organized around the theme of
paranormal claims. A good introduction to a skeptical view that includes short critiques of almost all popular fringe-science claims.

Scott, Eugenie C. 2001. "My Favorite Pseudoscience." In *Skeptical Odysseys:
Personal Accounts by the World's Leading Paranormal Inquirers*, edited by
Paul Kurtz. Amherst, NY: Prometheus.

Searle, John R. 1980. "Minds, Brains, and Programs." *Behavioral and Brain
Sciences* 3: 417–458. Introduces the Chinese Room thought experiment.

Seidman, Barry F., and Neil J. Murphy, eds. 2004. *Toward a New Political Humanism*. Amherst, NY: Prometheus. Gives a good sample of the political concerns of left-leaning nonbelievers as a minority interest group
in a time of global religious resurgence.

Shamos, Morris H. 1995. *The Myth of Scientific Literacy*. New Brunswick, NJ:
Rutgers University Press. About the difficulties of science education
and the impracticality of the ideal of having most people attain scientific literacy to the extent of truly grasping the interlocking structures
of the main theories in modern science.

Shanks, Niall. 2004. *God, the Devil, and Darwin: A Critique of Intelligent Design
Theory*. New York: Oxford University Press. An excellent introduction
to the biochemical and cosmological failures of intelligent design the-

ory, set in the context of the development of the philosophical argument from design to establish the existence of a God.

Shanks, Niall, and Istvan Karsai. 2004. "Self-Organization and the Origin of Complexity." In *Why Intelligent Design Fails: A Scientific Critique of the New Creationism*, edited by Matt Young and Taner Edis. New Brunswick, NJ: Rutgers University Press. Using examples such as Bénard cells and wasps' nests, Shanks and Karsai explain how the application of simple rules and self-organizing processes can lead to complex order.

Shermer, Michael. 2003. *How We Believe: Science, Skepticism, and the Search for God*. New York: Henry Holt. An accessible discussion of science, religion, the causes of religious belief and the reasons believers give for their faith. Especially interesting in presenting evidence that people most often give evidential and quasi-scientific reasons for their belief in a God.

————. 2004. *The Science of Good and Evil: Why People Cheat, Gossip, Care, Share, and Follow the Golden Rule*. New York: Times Books. Attempts to explain moral behavior within the natural world, and proposes a view of "provisional" morality that makes it clear how moral thinking is continuous with science. An accessible book, although sometimes at the price of oversimplification. Shermer's view of morality is unlikely to be convincing to those who do not already share his naturalistic views.

Smith, George H. 1989. *Atheism: The Case against God*. Buffalo, NY: Prometheus. One of the most popular books arguing that there is no God. Smith's case is entirely philosophical, with practically no reference to science. His case for atheism rests on his attempt to show that a particular conception of God, most apparent in conservative Catholic theology, does not make sense. Even if Smith were successful, his argument would do little to undermine alternative supernaturalistic conceptions of the world.

————. 1991. *Atheism, Ayn Rand, and Other Heresies*. Buffalo, NY: Prometheus. Essays defending atheism and the hypercapitalist philosophy of Ayn Rand. An example of right-wing libertarian nonbelief.

Smith, Huston. 2001. *Why Religion Matters: The Fate of the Human Spirit in an Age of Disbelief*. New York: HarperSanFrancisco. A profoundly antiscientific book by an influential scholar of religion, it illustrates the deep worries modern science can cause for devout people who are not willing to escape into the obscurantism of academic theology.

Smolin, Lee. 1997. *The Life of the Cosmos*. New York: Oxford University Press. Applies quasi-Darwinian population thinking to cosmology.

———. 2001. *Three Roads to Quantum Gravity*. New York: Basic. A nontechnical discussion of some current approaches to unifying gravity with quantum mechanics.

Sober, Elliott, and David Sloan Wilson. 1998. *Unto Others: The Evolution and Psychology of Unselfish Behavior*. Cambridge, MA: Harvard University Press. An evolutionary biologist and philosopher of biology explain how altruistic behavior and psychology emerge from biological evolution. Includes a somewhat controversial defense of group selection as a major factor influencing human behavior.

Sorabji, Richard. 1988. *Matter, Space and Motion: Theories in Antiquity and Their Sequel*. London: Duckworth. On the intimate connections between physics, philosophy, and theology in the ancient world.

Spanos, Nicholas. 1996. *Multiple Identities and False Memories: A Sociocognitive Perspective*. Washington, DC: American Psychological Association. Explains phenomena such as multiple personalities, false memories, UFO abductions, and spirit possession experiences.

Stanovich, Keith E. 2004. *The Robot's Rebellion: Finding Meaning in the Age of Darwin*. Chicago: The University of Chicago Press. A cognitive scientist presents an ambitious theory of rationality based on awareness that the reproductive interests of the genes and memes that make us up are not necessarily identical to those interests we identify with on rational reflection.

Stark, Rodney. 2004. "Fact or Fable? Digging Up the Truth in the Evolution Debate." *The American Enterprise* 15(6): 40–44.

Stark, Rodney, and Roger Finke. 2000. *Acts of Faith: Explaining the Human Side of Religion*. Berkeley: University of California Press. Sociologists of religion charge their field with adopting atheistic assumptions, claim scientists are as religious as anyone else, declare the secularization thesis dead, and argue that rational choice theory is the best approach for understanding modern religion.

Stein, Gordon. 1980. *An Anthology of Atheism and Rationalism*. Amherst, NY: Prometheus. A collection of historically important and representative essays concerning freethought and nonbelief. Not much from a scientific point of view appears, which is itself significant.

Stenger, Victor J. 2000. *Timeless Reality: Symmetry, Simplicity, and Multiple Universes*. Amherst, NY: Prometheus. A good semipopular resource to find out about developments in modern physics and cosmology concerning ideas such as symmetry breaking, time reversibility, and multiple universes.

————. 2003. *Has Science Found God? The Latest Results in the Search for Purpose in the Universe*. Amherst, NY: Prometheus. A physicist criticizes some common science-based arguments for the existence of a God. Especially strong in its discussion of cosmology and paranormal claims.

————. 2004. "Is The Universe Fine-Tuned for Us?" In *Why Intelligent Design Fails: A Scientific Critique of the New Creationism*, edited by Matt Young and Taner Edis. New Brunswick, NJ: Rutgers University Press.

————. 2006. *The Comprehensible Cosmos. Where Do the Laws of Physics Come From?* Amherst, NY: Prometheus. On how symmetry principles, notably point-of-view invariance, lead to the laws of physics.

Stoeber, Michael, and Hugo Meynell, eds. 1996. *Critical Reflections on the Paranormal*. Albany: SUNY Press. Essays by philosophers and theologians defending parapsychology and its potential to overcome naturalism and to support religion more directly by demonstrating that "agent causation" cannot be reduced to physical causes.

Strobel, Lee. 2004. *The Case for a Creator: A Journalist Investigates Scientific Evidence That Points toward God*. Grand Rapids, MI: Zondervan. A popular exposition of the ideas of evangelical scholars who argue that science provides evidence for God. Popular arguments for nonbelief that refer to science often develop as a reaction to these kinds of religious apologetics.

Sturrock, Peter A. 1999. *The UFO Enigma: A New Review of the Physical Evidence*. New York: Warner. One of the highest quality, intellectually serious books sympathetic to UFOs available. Although taking UFO claims seriously, the panel whose work led to this book can only manage a lukewarm recommendation for UFOs as a phenomenon worthy of further study.

Thrower, James. 2000. *Western Atheism: A Short History*. Amherst, NY: Prometheus. A short survey of atheism in the Western philosophical tradition. It also examines the early relationship of science and nonbelief.

Turner, James. 1985. *Without God, without Creed: The Origins of Unbelief in America*. Baltimore: The Johns Hopkins University Press. Particularly valuable in showing the similarities in attitude between some conservative

Christians and eventual nonbelievers due to a shared empiricist frame of mind.

Vitzthum, Richard C. 1995. *Materialism: An Affirmative History and Definition.* Amherst, NY: Prometheus. Explores and defends a materialist point of view, focusing on the historical examples of Lucretius, d'Holbach, and Büchner and bringing materialism up-to-date by examining twentieth-century versions. Highlights both the historical continuity of materialist ideas and the significant differences between ancient and modern materialism.

Ward, Mark. 2000. *Virtual Organisms: The Startling World of Artificial Life.* New York: St. Martin's Press. Written by a technology journalist, this is one of the most accessible introductions to artificial life research and the intellectual debates artificial life has inspired.

Warraq, Ibn. 1995. *Why I Am Not a Muslim.* Amherst, NY: Prometheus. One of the most prominent recent polemics by a nonbeliever of Muslim background. It includes a brief survey of Muslim philosophers, some of whom were quite skeptical regarding revealed religion. The author somewhat exaggerates the degree of doubt among Islamic philosophers.

Weinberg, Steven. 1992. *Dreams of a Final Theory: The Search for the Fundamental Laws of Nature.* New York: Pantheon. Accessible essays from a nonbelieving Nobel laureate physicist. Most relevant are two, "Two Cheers for Reductionism" and "What about God?"

————. 1999. "A Designer Universe?" *New York Review of Books*, October 21.

Wiebe, Donald. 1999. *The Politics of Religious Studies.* New York: St. Martin's Press. Criticizes the antiscientific, theological approach to religion prevalent in academic religious studies departments.

Wielenberg, Eric J. 2005. *Value and Virtue in a Godless Universe.* New York: Cambridge University Press. A critique of the notion that morality needs a God, and a defense of a naturalistic form of objective morality.

Wilber, Ken, ed. 1984. *Quantum Questions: Mystical Writings of the World's Great Physicists.* Boulder, CO: Shambhala. A collection of mystical and mystical-seeming writings from famous physicists responsible for developing quantum mechanics.

————. 1998. *The Essential Ken Wilber: An Introductory Reader.* Boston: Shambhala. In New Age circles, Wilber has acquired a reputation for profound, science-informed philosophy. This sampling of his typically opaque mystical prose might help with understanding why the New Age so often meets with contempt in mainstream scientific circles.

Wilson, David Sloan. 2002. *Darwin's Cathedral: Evolution, Religion, and the Nature of Society*. Chicago: The University of Chicago Press. Proposes to understand religion as a social glue that functions to keep societies together. Wilson portrays human societies as superorganisms shaped by group selection.

Wilson, Edward O. 1998. *Consilience: The Unity of Knowledge*. New York: Knopf. A prominent biologist's ambitious argument that all knowledge is continuous within a naturalistic, evolutionary framework. Includes a skeptical discussion of religion.

Windross, Tony. 2004. *The Thoughtful Guide to Faith*. New York: O Books. An Anglican minister, Windross defends an ultraliberal Christianity where the supernatural has become optional, science has the last word in describing the world, and in beliefs and attitudes, the modern Christian is practically no different from the secular humanist nonbeliever.

Witham, Larry A. 2002. *Where Darwin Meets the Bible: Creationists and Evolutionists in America*. New York: Oxford University Press. A journalist, Witham emphasizes the religious aspect of the debate over evolution in the United States. He interviews atheistic scientists, religious liberals who argue for separate spheres, and intelligent design proponents. A useful introduction to the religious and political aspects of the creation-evolution issue.

Woerlee, G. M. 2005. *Mortal Minds: The Biology of Near-Death Experiences*. Amherst, NY: Prometheus. An anesthesiologist explains how a brain near death generates often supernaturally interpreted experiences.

Wolpert, Lewis. 1993. *The Unnatural Nature of Science: Why Science Does Not Make (Common) Sense*. Cambridge, MA: Harvard University Press. One of the best discussions of the nature of science and how it consistently goes against common sense. Ordinary ways of thinking mislead us about science. Wolpert also addresses the conflicts between science and religious and paranormal beliefs.

Wright, J. Edward. 2000. *The Early History of Heaven*. New York: Oxford University Press. Contains interesting information on early Near Eastern and Ptolemaic conceptions of the universe, their effect on monotheistic scriptures, and the astral religions of antiquity.

Young, Matt. 2001. *No Sense of Obligation: Science and Religion in an Impersonal Universe*. Bloomington, IN: 1stBooks Library. A physicist argues that science describes a completely natural world, but also identifies

strongly with the Jewish tradition and explores ways of remaining religious without believing in a God.

Young, Matt, and Taner Edis, eds. 2004. *Why Intelligent Design Fails: A Scientific Critique of the New Creationism.* New Brunswick, NJ: Rutgers University Press. A team of scientists criticizes the intelligent design
version of anti-evolutionary thought in detail, arguing that scientifically it is a complete failure.

Zurek, Wojciech H. 1991. "Decoherence and the Transition from Quantum to
Classical." *Physics Today* 44(10): 36.

Zusne, Leonard, and Warren H. Jones. 1989. *Anomalistic Psychology: A Study
of Magical Thinking.* 2nd ed. Hillsdale, NJ: L. Erlbaum. A compendium
of mainstream psychological explanations of paranormal experiences
and beliefs.

Index

About the Author

TANER EDIS is Associate Professor of Physics at Truman State University. While primarily a theoretical physicist, he has also written extensively on the secularist tradition in science. He is the author of *The Ghost in the Universe: God in Light of Modern Science* (2002) and co-editor of *Why Intelligent Design Fails: A Scientific Critique of the New Creationism* (2004).